CALIFORNIA WATER CRISIS AND ITS IMPACTS: THE NEED FOR IMMEDIATE AND LONG–TERM SOLUTIONS

OVERSIGHT FIELD HEARING

BEFORE THE

COMMITTEE ON NATURAL RESOURCES
U.S. HOUSE OF REPRESENTATIVES

ONE HUNDRED THIRTEENTH CONGRESS

SECOND SESSION

Wednesday, March 19, 2014, in Fresno, California

Serial No. 113–63

Printed for the use of the Committee on Natural Resources

Available via the World Wide Web: http://www.fdsys.gov
or
Committee address: http://naturalresources.house.gov

U.S. GOVERNMENT PUBLISHING OFFICE

87–181 PDF WASHINGTON : 2015

For sale by the Superintendent of Documents, U.S. Government Publishing Office
Internet: bookstore.gpo.gov Phone: toll free (866) 512–1800; DC area (202) 512–1800
Fax: (202) 512–2104 Mail: Stop IDCC, Washington, DC 20402–0001

COMMITTEE ON NATURAL RESOURCES

DOC HASTINGS, WA, *Chairman*
PETER A. DEFAZIO, OR, *Ranking Democratic Member*

Don Young, AK
Louie Gohmert, TX
Rob Bishop, UT
Doug Lamborn, CO
Robert J. Wittman, VA
Paul C. Broun, GA
John Fleming, LA
Tom McClintock, CA
Glenn Thompson, PA
Cynthia M. Lummis, WY
Dan Benishek, MI
Jeff Duncan, SC
Scott R. Tipton, CO
Paul A. Gosar, AZ
Raúl R. Labrador, ID
Steve Southerland, II, FL
Bill Flores, TX
Jon Runyan, NJ
Markwayne Mullin, OK
Steve Daines, MT
Kevin Cramer, ND
Doug LaMalfa, CA
Jason T. Smith, MO
Vance M. McAllister, LA
Bradley Byrne, AL

Eni F. H. Faleomavaega, AS
Frank Pallone, Jr., NJ
Grace F. Napolitano, CA
Rush Holt, NJ
Raúl M. Grijalva, AZ
Madeleine Z. Bordallo, GU
Jim Costa, CA
Gregorio Kilili Camacho Sablan, CNMI
Niki Tsongas, MA
Pedro R. Pierluisi, PR
Colleen W. Hanabusa, HI
Tony Cárdenas, CA
Steven A. Horsford, NV
Jared Huffman, CA
Raul Ruiz, CA
Carol Shea-Porter, NH
Alan S. Lowenthal, CA
Joe Garcia, FL
Matt Cartwright, PA
Katherine M. Clark, MA

Todd Young, *Chief of Staff*
Lisa Pittman, *Chief Legislative Counsel*
Penny Dodge, *Democratic Staff Director*
David Watkins, *Democratic Chief Counsel*

CONTENTS

OVERSIGHT FIELD HEARING ON CALIFORNIA WATER CRISIS AND ITS IMPACTS: THE NEED FOR IMMEDIATE AND LONG–TERM SOLUTIONS

Wednesday, March 19, 2014
U.S. House of Representatives
Committee on Natural Resources
Fresno, California

The committee met, pursuant to call, at 10 a.m., in Fresno City Council Chambers, 2600 Fresno Street, Hon. Doc Hastings [Chairman of the committee] presiding.

Present: Representative Hastings, McClintock, Lummis, and Costa.

Also Present: Representatives Nunes, Valadao, Denham, and McCarthy.

The CHAIRMAN. The committee will come to order.

The House Committee on Natural Resources meets today to hear testimony on a hearing called the "California Water Crisis and Its Impacts: The Need for Immediate and Long-Term Solutions."

By way of introduction, I am Congressman Doc Hastings, and I have the privilege of representing the Fourth District in Washington State, which is very similar to the valleys here in California, although not as large. We have a tremendous amount of irrigated agriculture in my area, and I was down here 3 years ago when we had a subcommittee hearing. So, it is nice to be back in Fresno.

By way of introduction, before we get started, I just wanted to let you know, for those of you who aren't familiar as to how committee hearings are set, at least in my committee in the U.S. House, it is set up exactly this way, meaning that Democrats sit to my left and Republicans sit to my right. That is the way it is set up, and so I just wanted to let you know.

I should also note, too, that I sent a letter out to all members of the California delegation on both sides of the aisle to attend this hearing because we thought it was that important.

So, with that, there are a number of Members that are members of the House Committee on Natural Resources, but there are some Members that are not. So this is the procedure that we have to go through, and I ask unanimous consent that Mr. Nunes and Mr. McCarthy, Mr. Denham, who is on his way, and Mr. Valadao be allowed to sit at the committee, and without objection, they will be seated at the committee.

And with that, let me turn it over to my good friend and colleague from this area, Mr. Valadao, for the purposes of the opening ceremonies.

Mr. Valadao, you are recognized.

Mr. VALADAO. Thank you, Chairman Hastings. Thank you for taking the time to come out and visit us here in the valley. This is obviously a very important hearing, and this issue has been something that a lot of Members have been fighting before my time in Congress, and I am lucky enough to be able to join them in this fight this past year as I have served in the House of Representatives.

Being that this is a congressional hearing, we are going to begin, as we do with every session of the House of Representatives, with a prayer, the posting of the colors, and the Pledge of Allegiance.

So I would like to recognize Reverend Gaspar Baptista to offer a prayer.

[Prayer.]

Mr. VALADAO. Thank you, Father.

Please stay standing.

I would like to recognize Selma High School Marine Corps Junior ROTC to present the colors.

[Colors presented.]

Mr. VALADAO. As a token of our appreciation, I would like to present the Selma High School Marine Corps ROTC with a flag that was flown over the capital for their service today.

And to do our pledge, I want to invite my friend, Mr. William Bordeaux, retired U.S. Marine Corps veteran, to lead us in the Pledge of Allegiance.

[Pledge of Allegiance.]

The CHAIRMAN. The procedure that we will follow now will be that each Member will have up to 5 minutes to make opening remarks, and then we will go to our panel to hear their testimony. Following that, each Member will have some time to follow up on questions with any of the panelists that they desire.

So, with that, I will recognize myself for 5 minutes for my opening statement.

STATEMENT OF THE HON. DOC HASTINGS, A REPRESENTATIVE IN CONGRESS FROM THE STATE OF WASHINGTON

The CHAIRMAN. As I mentioned, it is a pleasure for me to be back here in Fresno because my regions that I have the privilege to represent are very much the same as the Central Valleys of California in that they are deserts, but they were transformed into the most productive soils on earth after our Nation wisely realized that the deserts would bloom if they had water and the necessary infrastructure of multi-purpose dams, reservoirs and canals.

Entire agriculture economies and ways of life grew around our ability to irrigate these once-dormant lands. In our regions, respective regions, food grows where water flows. Conversely, communities wither and die when the water spigot stops. As many of you know, this is an all-too-familiar occurrence here in the San Joaquin Valley. And, if it can happen here, it can happen anywhere in the Western United States, including my home area.

As Chairman of the House Natural Resources Committee, I have painfully watched the cycle that has gone on over here for the past several years. In 2009 and 2010, a man-made drought brought this region to its knees, where communities experienced 40 percent unemployment and food lines handed out Chinese-produced carrots to the victims of senseless regulatory drought. That followed, then, with two good years of water.

During those relatively good times, the House passed a comprehensive bill—and, I might add, every Member on this dais voted for that bill—that was intended to ensure that man-made drought never returned. Unfortunately, that bill fell on deaf ears in the Senate and by this administration. And so now we are back yet with another drought, and this one could be far more catastrophic than the previous one. History is once again repeating itself.

It doesn't have to be that way. As a matter of fact, had the bill that we passed out of the last Congress passed and been signed into law, we wouldn't be in this particular situation today because of the solutions that were embodied in that law.

But in the long term, we have an opportunity to right the wrongs of what has gone on here and to learn lessons from the past. First, let's stop the deliberate diversion of billions of gallons of water to a 3-inch fish where the science has not demonstrated that the fish is being helped by diversions. When the Ninth Circuit Court of Appeals validated the delta smelt Biological Opinion last week, it said that the species have been, and I quote, "afforded the highest of priorities by law even if it means the sacrifices of the anticipated benefits of the project." The opinion went on to say, and I quote again, "Resolution of this fundamental policy question lies with the Congress." I concur, and that is why we are here today. The Endangered Species Act needs reform, and we plan on our committee to begin that process in the next several weeks.

Second, let's reinvest in new water storage to capture water in wet times so that we can deliver it in dry times. The West is what it is today because of storage projects, and we are literally throwing that legacy away by thinking that conservation alone can resolve water shortages, because it can't.

There is a demonstrated willingness from non-Federal interests to invest in storage. The problem is that a maze of Federal Government regulations, including the Biological Opinion of the delta smelt, will continue to make it a difficult process for these projects to become a reality. I note that feasibility studies on three storage projects here in California have gone on for over a decade. I would just remind you, our country put a man on the moon in less time than that. This administration must stop getting ready to get ready for new storage. The California water bill that recently passed in the last Congress and this Congress immediately authorizes construction of storage by eliminating bureaucratic paralysis-by-analysis and by tapping into private investments.

And third, parties on all sides must have the political will to find common-sense solutions. To only resolve this situation in the short term is simply kicking the can down the road. I only wish that more of our California colleagues were here today. As I mentioned, I had invited them to be here today.

I also want to say that Mr. Valadao's recently passed bill that we passed, which is similar to the bill that passed last time, is a long-term solution to our problems that we face. And I might add again that all Members sitting in front of you voted for that legislation.

So now it is truly the time for the Senate to act. We cannot enact law in our country unless both houses act. We have acted in two Congresses in a row. It is now time for the Senate to act.

[The prepared statement of Mr. Hastings follows:]

PREPARED STATEMENT OF THE HONORABLE DOC HASTINGS, CHAIRMAN, COMMITTEE ON NATURAL RESOURCES

It's always a pleasure to be back in Fresno.

Our regions have a lot in common. They were once deserts, but they were transformed into the most productive soils on earth after our Nation wisely realized that the deserts would bloom if they had water and the necessary infrastructure of multipurpose dams, reservoirs and canals. Entire agricultural economies and ways of life grew around our ability to irrigate these dormant lands.

In our regions, food grows where water flows. Conversely, communities wither and can die when the water spigot stops. As many of you know, this is an all-too-familiar occurrence here in the San Joaquin Valley. And, if it can happen here, it can happen anywhere in the Western United States.

As Chairman of the House Natural Resources Committee, I have painfully watched the cycle that has gone on here over the past few years. In 2009 and 2010, a man-made drought brought this region to its knees, where communities experienced 40 percent unemployment and food lines handed out Chinese-produced carrots to the victims of senseless regulatory drought. This followed with 2 good water years.

During those relatively good times, the House passed a comprehensive bill intended to ensure that man-made drought never returned. That bill fell on deaf ears in the Senate and the administration. Now, we are back to yet another drought and this one could be far more catastrophic than before. History is once again repeating itself.

It doesn't have to be this way. We have an opportunity to right the wrongs of what's gone on here and to learn lessons from the past.

First, let's stop the deliberate diversion of billions of gallons of water to a 3-inch fish when the science has not demonstrated that the fish is being helped by the diversions. When the Ninth Circuit Court of Appeals validated the delta smelt Biological Opinion last week, it said that species have been quote "afforded the highest of priorities by law even it means the sacrifices of the anticipated benefits of the project" unquote. The opinion went on to say that quote "Resolution of these fundamental policy questions . . . lies with Congress." Unquote. I concur. The Endangered Species Act needs reformed and we plan to begin that process in the next few weeks.

Second, let's re-invest in new water storage to capture water in wet times so we can deliver it in dry times. The West is what it is today because of storage projects and we are literally throwing that legacy away by thinking that conservation alone can resolve water shortages.

There is a demonstrated willingness from non-Federal interests to invest in storage. The problem is that a maze of Federal environmental regulations—including a Biological Opinion on delta smelt—will continue to make it a difficult process to make these project a reality. I note that feasibility studies on three storage projects here in California have gone on for over a decade. Our great country put a man on the moon in less time. This administration must stop getting ready to get ready on new storage. The California water bill recently passed by the House immediately authorizes construction of storage by eliminating bureaucratic paralyses-by-analyses and by tapping into private investment.

Third, parties on all sides must have the political will to find common-sense solutions. To only resolve this situation in the short-term is simply kicking the can down the road. I only wish that more California House and U.S. Senate Democrats, all of which I invited, could be sitting here to listen to the long-term solutions posed today.

Mr. Valadao's recently passed bill is a great start to bringing short and long-term help. Long-term solutions must be part of the equation or we will repeat this destructive drought cycle once again. We must not be short-sighted. I hope that the Senate can do its part so a meaningful but rapid negotiation can truly happen. The Senate is entitled to its solution, but the next step is to pass their solution and if there are differences, get together to work differences out.

In closing, I want to thank Mr. Valadao for his leadership on his bill and in asking for this hearing. The people of the San Joaquin Valley have asked for and deserve nothing less than their water. We are here to listen and to bring your message back to Washington, DC.

With that, I recognize the distinguished Member who is serving as the Ranking Member of this committee, Mr. Costa, for 5 minutes.

STATEMENT OF THE HON. JIM COSTA, A REPRESENTATIVE IN CONGRESS FROM THE STATE OF CALIFORNIA

Mr. COSTA. Thank you very much, Chairman Doc Hastings, for holding this hearing, and my colleagues for being here today.

Obviously, this devastating drought is impacting our valley, but I think it is reflective of the water problems our entire State is going to have. No people, though, will bear the brunt of this drought's effects more than the people in this room today and many who are outside of this room who are part of the farmers, the farm workers, and the farm communities that have made our valley what it is today.

We see its effects in lost jobs, in families standing in food lines to provide for their meals, and in the incredible burden that mothers and fathers have trying to provide the basic needs for their families while fertile ground beneath their feet lies dry and fallow.

I believe that this drought and the comments that I have made reflect the sentiment that every person in this valley has. And like many of you, I am angry. I am angry that we have failed to capture the water that lies available to us in times of plenty so that we can use it in years like these. I am angry that we have failed to invest in our water system that was designed by our parents and our grandparents. And unfortunately today, California's needs have become over-burdened because of the greater population increase, and the water supply is stretched thin, and therefore we have a broken water system.

I am angry that in face of the devastation, we continue to point fingers and play the blame game, and it does not bring us one additional drop of water.

To understand my anger, we must first place this drought in perspective. Like many of you, my family and I today have farmed for generations. We went through this from 2008 to 2010, as the Chairman noted. It was terrible. From the drought, we learned many things, but the biggest lesson we should have learned is that we can simply not continue to do nothing. We must act and we must act now to resolve this problem, and we must do it in a bipartisan fashion. We must forge compromises that ultimately result in successful legislation. That is the only way you ever get successful legislation.

Every day, I work with everyone to bring more water to our valley, and no one is more committed to doing that than the folks that are here on this dais. I certainly am. And we have had success when we work together: the Intertie Project; water bonds that in the past and today are being considered for additional storage in Sacramento; carryover storage. Just this week, the State board order has been changed to relieve the pressure on the exchange contractors and other water users.

In 2011, I introduced H.R. 1251, more water for our valley after the 2010 drought which provided more direction on how the pumps should operate while still being in compliance with the Endangered Species Act. Now, I think we ought to change the Endangered Species Act, but I also believe that the likelihood of that occurring in the Senate is not good.

Had this bill been in place, though, it is estimated that we would not have lost the 800,000 acre feet of water last year. Unfortunately, that bill was never brought for a hearing, let alone a vote.

But when we work together, we can have success. The most recent success was when the colleagues, many here at this table, pushed the Bureau of Reclamation and Director Murillo that the carryover water that had been saved last year by farmers, 360,000 acre feet of water, must be made available for those farmers who saved that water. That was a success.

But we need to have more of this type of cooperation to craft legislation that can successfully pass in the House of Representatives as well as in the U.S. Senate that could be signed by the President.

We all know what the solutions must include: more storage in the long term so we can save water in wet years; an improved conveyance system that doesn't cause the collapse of the delta so that we can move water through the Sacramento San Joaquin River Delta system; and legislative flexibility, legislative flexibility to provide the operations under the State and Federal projects. That is why this year I have introduced and supported bills to provide authorization of more storage. Congressman McClintock, who chairs the subcommittee, he has introduced H.R. 937, and I am a co-sponsor of that to raise the spillway gates at Exchequer Dam that would provide 70,000 acre feet of additional water. We passed that bill in the House; it is now in the Senate. We should act on it. H.R. 4125, which would authorize the expansion of Exchequer Dam that I introduced last month, 600,000 additional acre feet of water; H.R. 4126, which would enlarge San Luis Reservoir in my district that would provide 130,000 acre feet of additional water; and Temperance Flat Dam that we share here that would provide us the ability to move water north and south that would provide 1.3 million acre feet of additional storage.

However, we must be able to supply more water for our farms and for our cities. And as I close, I want to suggest that legislation that has been introduced this year, H.R. 4039, would provide direction on how we can operate the systems that capture water that we need so much.

In closing, let me say that none of us can make it rain. If we could, we would have done it already. But our valley is our home. We love it dearly. And my heart goes out to all of those who have been impacted, as my family and your family has. I look forward to listening to the testimony, Mr. Chairman, of the witnesses, and continuing to work with you and members of the committee to solve our drastic water drought that is facing the people of this valley and California.

The CHAIRMAN. I thank the gentleman for his statement.

I recognize the gentleman, another gentleman—in fact, I am going to recognize every gentleman from California that I will recognize will be from California.

[Laughter.]

The CHAIRMAN. That will not be the case. After Mr. McClintock, it will be somebody from Wyoming.

So I will recognize the gentleman from California, Mr. McClintock.

STATEMENT OF THE HON. TOM McCLINTOCK, A REPRESENTATIVE IN CONGRESS FROM THE STATE OF CALIFORNIA

Mr. McCLINTOCK. Thank you, Mr. Chairman.

As you know, long before the current drought, the Central Valley suffered from the deliberate diversion of billions of gallons of water promised to it under the Bay Delta Accord. Instead, that water was dumped into the Pacific Ocean for the amusement of the delta smelt. This wanton act caused the loss of a quarter-million acres of the most productive farmland in America. It threw thousands of families into unemployment. It economically devastated this region.

Well, now a natural drought has compounded the regulatory drought, and here is the simple truth of the matter. Droughts are nature's fault; water shortages are our fault. Nature produces 45,000 gallons of fresh water every day for every man, woman and child on this planet. The problem is that water is unevenly distributed over both time and distance. We build dams to transfer water from wet times to dry times. We build aqueducts to transfer water from wet areas to dry areas. We don't build dams and aqueducts to dump that water into the ocean.

[Applause.]

Mr. McCLINTOCK. Water is very good at flowing downhill on its own. It doesn't need our help. We build dams and aqueducts so that surplus water isn't lost to the ocean but is rather retained for human prosperity.

Unfortunately, in the last generation, a radical and retrograde ideology has insinuated its way into our public policy. It holds that human needs need to be subordinated to the goal of restoring Earth to her prehistoric condition. In pursuit of this goal, this movement has obstructed the construction of new dams by attaching so many conditions and restrictions as to render them economically infeasible. Mr. Costa just referenced bills that ostensibly authorize new dams at sites and Temperance Flats, but only if judged feasible under these unobtainable standards. Translation, that means these dams will not get built.

We have been unable to get the spillway raised just 10 lousy feet on the Exchequer Dam that would add 70,000 acre feet of additional storage to Lake McClure. Self-described environmentalists oppose it because it would require a minor boundary adjustment to the wild and scenic river boundary that overlapped with the pre-existing FERC boundary. Indeed, this movement has not only obstructed the construction of new dams, it has actively pursued the goal of tearing down existing ones such as the four hydro-electric dams on the Klamath.

If anything good comes of this drought, it will be that the public is finally awakening to the enormous economic and environmental damage that these policies have done.

The House has acted twice on legislation to address both the regulatory drought caused by the unnecessary water diversions and to begin removing the regulatory hurdles that block new dam construction. Mr. Valadao's H.R. 3964 is an important first step. It strengthens water rights, it stops the massive loss of water required by the Biological Opinions on smelt in the San Joaquin River Restoration Act, it opens up additional storage for local agencies at New Melones, it expands capacity at Lake McClure, it allows local water districts to partner with the Federal Government to expedite expansion and construction of reservoirs.

But there is a problem, and the problem is that the Senate has not acted on this legislation, and progress cannot be made between the two houses until the Senate either passes the House bill or sends its own bill to the House so that the conference process can proceed to a conclusion.

We are at a crossroads, and it is time to choose between two very different visions of water policy. One is the nihilistic vision of increasingly severe government-induced shortages, higher and higher electricity and water prices, massive taxpayer subsidies to politically well-connected industries, and a permanently declining quality of life for our children who will be required to stretch and ration every drop of water and every watt of electricity in their bleak and dimly lit homes.

The other is a vision of abundance, a new era of clean, cheap, and abundant hydro-electricity, great new reservoirs to store water in wet years, to assure abundance in dry ones, a future in which families can enjoy the prosperity that abundant water and electricity provide and the quality of life that comes from that prosperity. It is a society whose children can look forward to a green lawn, a backyard garden, affordable air-conditioning in the summer and heating in the winter, brightly lit homes and cities, and abundant and affordable groceries from America's agricultural cornucopia.

I yield back.

[Applause.]

The CHAIRMAN. I thank the gentleman for his opening statement.

I now recognize my colleague and a member of the Natural Resources Committee from Wyoming, Mrs. Cynthia Lummis, for 5 minutes.

STATEMENT OF THE HON. CYNTHIA M. LUMMIS, A REPRESENTATIVE IN CONGRESS FROM THE STATE OF WYOMING

Mrs. LUMMIS. Thank you, Mr. Chairman, and thanks to my colleagues from California for inviting me here to discuss the dire water situation in the San Joaquin Valley.

What is happening to the people of the San Joaquin Valley strikes at the core of what western communities share in common. Like you, water is our lifeblood in Wyoming. Without it, people suffer. People are suffering in California right now because of the water shortage, job losses, unemployment, and fallowed farmland.

I say with great regret that some of the factors contributing to the water shortages are man-made. The Congress can't control the weather, but the Congress can act to strike a responsible balance between the needs of people and the needs of fish, instead of letting taxpayer-funded lawsuits determine the fate of the San Joaquin Valley.

The livelihoods of people are being held hostage by groups who sue over and over to get their way whenever Fish and Wildlife officials disagree with them. After hundreds of millions of dollars spent on the delta smelt, there is still no scientific consensus as to why it is declining. Nonetheless, using the courts and tax dollars to pay their attorney fees, environmental litigants like the NRDC have managed to choke off water supplies to the valley, all for uncertain benefit to the fish.

Together with Chairman Hastings and Representative Valadao, we formed a working group to study the Endangered Species Act and examples like this where the law simply has not worked as intended. We will be pursuing legislation in the months ahead to improve the law and make it work better for both species and people. The hope is to re-focus the law on actually recovering species and prevent it from being abused like it has been here in the San Joaquin Valley.

But in the meantime, the people in the valley need immediate relief. That is why the House of Representatives passed Mr. Valadao's bill, H.R. 3964. The legislation is designed to restore long-term water supplies to farmers and communities that right now are looking at zero or near-zero percent allocations of water. The bill will also promote more water storage to meet the needs of California's growing population, and of farmers who are feeding more than half our Nation with vegetables, fruits and nuts. California's agricultural community has taken water conservation to a new level over the last decades. They have never been more efficient, and their techniques are being duplicated across the country.

But when you have a water storage and delivery system built for 22 million people and a population of 38 million people, and growing, no amount of water conservation is going to completely solve your problem. California and the people of the San Joaquin Valley need more water storage, period. These are the good people of agriculture who perform a noble task, who feed their families and mine, who love and nurture the land and the culture of our Nation.

I am here from Wyoming today to honor these families and to support their goal of continuing this honorable and important work of providing food and fiber, the cornerstones of our Nation's security, to all Americans.

Mr. Chairman, that is enough for me. The real story will be told by the impressive panel of California witnesses before us. I thank you, Mr. Chairman, and I thank the Members of the California here today, both on the dais and in attendance, and I yield back.

[Applause.]

The CHAIRMAN. I thank the gentle lady for her testimony.

Now I will recognize somebody that you are all very familiar with, Mr. Nunes from California.

Mr. Nunes.

STATEMENT OF THE HON. DEVIN NUNES, A REPRESENTATIVE IN CONGRESS FROM THE STATE OF CALIFORNIA

Mr. NUNES. Thank you, Mr. Chairman, and thank you for your continued pursuit of policies that make sense in the Congress.

Mr. Chairman, the environmental lobby and its allies attribute the current water crisis to historic drought induced by global warming, but they have it wrong. Central Valley farmers are fallowing their lands not because of low levels of rainfall, our communities are restricting water usage not because the world is getting warmer, and our water districts are receiving a fraction of their allocations but not because of greenhouse gases. The problem is not that the Central Valley is lacking water, it is that we are being deprived of it.

Our forefathers blessed this State with an amazing irrigation system that could withstand 5 years of drought. The principle was simple: water was captured during the wet years for use in the dry ones. We are still capable of doing that, but we are not allowed. For decades, preposterous environmental regulations have put more and more water off limits to people. Before these regulations began taking hold in the early 1990s, water districts routinely received 100 percent of their water allocations. Since then, they have hardly ever received 100 percent. And today, this year, many will receive zero.

For the sake of supposedly persecuted fish species, the salmon and the most precious of all, the 3-inch bait fish called the delta smelt, government authorities have diverted enormous supplies of water from human usage. How much water has been lost? Consider this: in the past 7 years, nearly 4 million acre feet of water that could have been used by families and farmers has been flushed into the ocean. That is not a result of global warming or drought. It is the result of a government run amok.

Altogether, the diversion of water from farms and communities to environmental causes has created an average annual water shortfall of roughly 1.75 million acre feet on the west side of this valley, and 250,000 acre feet on the east side of the valley, plus an additional east side groundwater overdraft of 400,000 acre feet. That leads to a total average shortfall of nearly 2.5 million acre feet.

What does this mean? If nothing is done, around 800,000 acres of productive farmland will be forced permanently out of production.

The House of Representatives passed a bill 6 weeks ago and a similar bill in 2012 that would eliminate this shortfall completely. If the new bill had been in effect for the last 7 years, about half the water flushed into the ocean would have been kept for use.

More recently, after years of total inaction on this grave problem which destroyed thousands of jobs for farmers and farm workers, the Senate finally introduced a bill that would alleviate some of the water shortfall. It would supply around 300,000 acre feet. This leaves a deficit of 2.1 million acre feet.

Of course, simply introducing a bill doesn't achieve anything. The Senate needs to pass it so that the House and Senate can go to conference and hammer out a compromise bill. But it must be crystal clear to everyone: if a compromise bill provides anything less than 2.4 million acre feet, then the State of California counties, cities and irrigation districts, not the Federal Government, will be responsible for overcoming the remaining deficit.

What many people don't realize is that the fallowing of farmland and the uprooting of entire communities is not a side-effect of the radical environmentalists' actions. It is their goal. For extremists who view human settlement and productive economic activity as a blight on nature, it is a good thing that water-deprived families are abandoning their farms and homes. These extremists won't uproot themselves from their comfortable homes in San Francisco and other coastal cities, but they are more than willing to use the Central Valley communities as a guinea pig to see if our lands can be restored to some mystical state of nature.

These radicals never tire of battle and they never give up. You see the results of their relentless fight in the new dustbowl that has overtaken the valley. They are using all their influence to oppose the House-passed water bill that would permanently end the crisis, and they recently imposed their will on the Metropolitan Water District of southern California.

Meanwhile, many of our Ag groups' lawyers and lobbyists that are supposed to be representing the valley are acting to protect their own self-interests with high-paying jobs, pensions, and unwillingness to actually call a spade a spade. When a government can't provide water to its citizens, the government has failed. Victims of its policies have two options, to rise in protest until these policies are changed or watch helplessly as their communities are destroyed. The leftists who have engineered this crisis understand this. To try to keep their victims from protesting, they have offered up a scapegoat of global warming, and to buy them off, they have offered programs and bills that don't provide water but do provide sums of money, essentially giving welfare to people who really only want water.

By passing two comprehensive bills, the House of Representatives has already shown that it is on the side of the people of the valley, but that is not enough to restore their water. As shown by the Ninth Circuit Court last week in favor of the delta smelt, even during an historic water crisis, the radical environmental agenda advances remorselessly. The smelt may have won another victory, but the struggling farmers and thirsty families and shattered communities of this valley are paying the price.

I have been fighting these extremists for more than a decade in Congress, and as long as I am elected I will continue to keep up this fight. The most important thing is to give a voice to the people who will not sit idly by while their livelihoods are stolen away from them and the hopes and dreams of their children and grandchildren are grinded into dust.

I yield back.

[Applause.]

[The prepared statement of Mr. Nunes follows:]

Prepared Statement of The Honorable Devin Nunes, a Representative in Congress From the State of California

The environmental lobby and its allies attribute the current water crisis to a historic drought induced by global warming. But they have it wrong. Central Valley farmers are fallowing their lands, but not because of low levels of rainfall. Our communities are restricting water usage, but not because the world is getting warmer. And our water districts are receiving a fraction of their allocations, but not because of greenhouse gases.

The problem is not that the Central Valley is lacking water, it's that we're being deprived of it.

Our forefathers blessed this State with an amazing irrigation system that could withstand 5 years of drought. The principle was simple—water was captured during wet years for use in dry ones.

We're still capable of doing that, but we're not allowed. For decades, preposterous environmental regulations have put more and more water off-limits to people. Before these regulations began taking hold in the early 1990s, water districts routinely received 100 percent of their water allocations. Since then, they've hardly ever received 100 percent, and today many actually receive 0 percent.

For the sake of supposedly persecuted fish species—the salmon, the steelhead, and the most precious of all, the 3-inch baitfish called the delta smelt—government authorities have diverted enormous supplies of water from human usage. How much water has been lost? Consider this: in the past 7 years, 3.9 million acre feet of water that could have been used by families and farmers have been flushed out into the ocean. That is not a result of global warming or drought—it's a result of government run amuck.

Altogether, the diversion of water from farms and communities to environmental causes has created an average annual water shortfall of 1.75 million feet on the westside and 250,000 acre feet on the eastside, plus an eastside groundwater overdraft of 400,000 acre feet. That leads to a total average annual shortfall of 2.4 million acre feet in the Central Valley. This means if nothing is done, around 800,000 acres of productive farmland will be forced out of production.

The House of Representatives passed a bill 6 weeks ago, which followed a similar bill passed in 2012, that would eliminate this shortfall completely. If the new bill had been in effect for the last 7 years, about half the water flushed into the ocean would have been kept in usage—and some might say we could use an extra 2 million acre feet of water right now.

More recently, after years of total inaction as this grave problem destroyed thousands of jobs for farmers and farm workers, the Senate finally introduced a bill that would alleviate some of the water shortfall—it would supply around 300,000 acre feet, leaving a deficit of 2.1 million acre feet. Of course, simply introducing a bill doesn't achieve anything—the Senate needs to pass it, so that the House and Senate can go to conference and hammer out a compromise bill.

It must be crystal clear to everyone, however, that if a compromise bill provides anything less than the 2.4 million missing acre feet, then the State of California, counties, cities, and irrigation districts—not the Federal Government—will be responsible for overcoming the remaining deficit.

What many people don't realize is that the fallowing of farmland and the uprooting of entire communities is not a side-effect of the radical environmentalists' actions; it is their goal. For extremists who view human settlement and productive economic activity as a blight on nature, it's a good thing that water-deprived families are abandoning their farms and homes. These extremists won't uproot themselves from their comfortable houses in San Francisco and other coastal cities, but they're more than willing to use Central Valley communities as guinea pigs to see if our lands can be restored to some mystical state of nature.

These radicals never tire of the battle, and they never give up. You see the results of their relentless fight in the new dustbowl that has overtaken the Central Valley. They are using all their influence to oppose the House-passed water bill that would permanently end this crisis, and they recently imposed their will on the Metropolitan Water District of Southern California. Meanwhile, Ag groups, lawyers, and lobbyists who are supposed to be representing valley farmers are primarily acting to protect their own high-paying jobs and pensions.

When a government cannot provide water to its citizens, that government has failed. Victims of its policies have two options—to rise up and protest until these policies are changed, or watch helplessly as their communities are destroyed. The leftists who engineered this crisis understand this. To try to keep their victims from protesting, they have offered up the scapegoat of global warming. And to buy them off, they have offered programs and bills that don't provide water, but do pay out large sums of money, essentially giving welfare to people who just want to work.

By passing two comprehensive water bills, the House of Representatives has already shown that it's on the side of the people of the Central Valley. But that is not enough to restore their water. As shown by the Ninth Circuit Court's ruling last week in favor of the delta smelt, even during a historic water crisis, the radical environmental agenda advances remorselessly. The smelt may have won another victory, but the struggling farmers, the thirsty families, and the shattered communities of the Central Valley are paying the price.

I have been fighting these extremists for more than a decade in Congress, and as long as the people of this valley entrust me as their representative, I will continue fighting. I will give a voice to people who will not sit idly by while their livelihoods are stolen away from them and the hopes and dreams of their children and grandchildren are grinded into dust.

The CHAIRMAN. I thank the gentleman for his statement.

I will now recognize the majority whip from Bakersfield, Mr. McCarthy.

STATEMENT OF THE HON. KEVIN McCARTHY, A REPRESENTATIVE IN CONGRESS FROM THE STATE OF CALIFORNIA

Mr. McCARTHY. Thank you, Mr. Chairman.

First, thank you all for turning out. And I want to thank the committee for being here. You know, this isn't the first time the committee is here. We were here 3 years ago, we were here 4 years ago, because we knew today would come, and we wanted to be better prepared.

The sad part is the House has done the work. The Senate just stood there defensively in the face of the drought. I am here to tell you that is not a noble gesture.

Just yesterday I was reading *The Wall Street Journal*, headline, front page. Because of the drought in California and Texas, prices are rising. We know from the valley what we grow is already affecting the economy. We see it right here, right now.

Where will we get our food? Do you ever question about food safety? There is no better place to do it than grow our own. There is no better place to grow it than here.

The challenge is, do we have to be where we are today? We know some years will be dryer than others. On record, this is one of the driest. Three years ago we had 170 percent of snow pack, just 3 years ago. What was the State allocation then? Eighty percent. Eighty percent in a year of 170.

Devin has led this charge when many people thought it was just a valley issue. He has been able to grow it into a national issue, because it is. The policies that are affecting this Nation aren't just going to affect California but will affect the entire country, and it holds us back.

David has led the charge this time in Congress to actually pass legislation. The Senate has done nothing.

The challenge is, in 1994, something unique happened. Republicans, Democrats, environmentalists, water users all got together and made an agreement. They made an agreement, and in that agreement it said in the wet years we would let water come down from the delta. We would be able to store it for years like today. But court cases, Federal regulation and others have made that different, so we are in a different place.

My wife and I, we have two children. Like you, our greatest fear in the future is what the future will look like for them. What are the opportunities that they will have? And when they are growing up, you want to teach them lessons of history and the past, and you study history. Are there tales, are there fables you can teach them to teach them the lessons?

Have any of you ever studied Aesop, the ancient Greek? He was a slave. He would tell these fables to teach a lesson for a moral history in the future. And there is a famous one that we would always—we would read it and he would tell it—about the ant and the grasshopper. It is really quite simple. It is a little story of an ant and a grasshopper in the summertime. What did the ant do? The ant went and stored the food, took it down, afraid that there might be a cold winter, when others would say, oh, no, nothing will happen. The grasshopper enjoyed the summer, afraid nothing ever will come.

Well, you know what? A cold winter came. Who was prepared?

This would have been a great year to be an ant. We had a great opportunity. But government is pushing us into being grasshoppers, a challenge we do not have to be.

Our Chairman is from Washington State. Earlier this year they were at 60 percent of their snow pack. The President came out here, brought our two Senators, to talk about the water crisis of California. He thinks it is a different issue. In that time we had that little bit of rain, Washington had 40 percent more of their snow pack. I think the President is on the wrong path that he thinks it is just global warming. Washington is not far from California.

But do you know, during that little bit of rain, in these type of years it is very important. I just heard a report back that projected from a water district that 445,000 acre of feet went out to the ocean because of State and Federal regulation. State quality says it needs to be 8-to-1. The fish get 8; we get 1.

It kind of goes back to the old fable. It is not that difficult. We need more storage, we need the ability for the water to come down, not into the ocean, and we need a plan for the years ahead. We were here 3 years ago to make that plan. We were very fortunate in a process that most States don't have. Both our Senators are committee chairs. Do you know how powerful it is to be a committee chair in the Senate? But there still has not been a vote in the Senate on water legislation for California.

This country was created with a House and a Senate. One House passes a bill, the Senate passes a bill, and you go to conference and you solve the problem. We have been waiting two Congresses now. We have gotten the entire Congress together to vote on where to move, just focused on California. Very seldom does that ever happen. The question now is, will the Senate even act?

We can't wait any longer. We are now in that place that we did not want to be, and I believe there is a better opportunity. I look forward to hearing from all, and I yield back.

[Applause.]

The CHAIRMAN. I thank the gentleman for his testimony.

Now I will recognize another representative from the Central Valley. Mr. Denham is recognized for 5 minutes.

STATEMENT OF THE HON. JEFF DENHAM, A REPRESENTATIVE IN CONGRESS FROM THE STATE OF CALIFORNIA

Mr. DENHAM. Thank you, Mr. Chairman. Thank you for holding yet another hearing here in the valley, and certainly one more of many hearings that we have held in Washington, DC on our water crisis.

I start by apologizing for being late, coming from the northern part of the valley. If we had been spending the last 5 years' stimulus dollars on shovel-ready transportation projects rather than high-speed rail, I would have been here earlier.

[Laughter.]

[Applause.]

Mr. DENHAM. And I only bring up high-speed rail in this context in a water hearing. If California had its priorities right, we would be building water storage instead of a train that may never come.

[Applause.]

Mr. DENHAM. Two years ago, the House passed legislation seeking to address the imbalance in the State's water system, and it was only met with criticism and dismissal. Criticism from our Senate counterparts was met with no solutions and no legislative proposals at the time.

Now the State finds itself in the midst of a drought. The Speaker of the House came to the valley to witness the devastation, and the House again acted swiftly.

The President visited recently, bringing news of help for food banks and discussing the ills of climate change.

Even the Senate finally placed ideas into legislative text, but the scope is short-term and focuses on the authority regulatory agencies already possess.

Unfortunately, the President needs to ensure we don't end up in this great crisis once again.

A bigger problem facing the valley is the outlandish approach the California State Water Resources Control Board has taken in the last year on required flows for the Tuolumne River, and most recently with the threat of overtaking the Central Valley project and throwing out over 100 years of water rights history.

Added to this list is the State board's most recent idea of cutting off all of agriculture for 2 years.

Over a year ago, the State board proposed to send over 35 percent of unimpaired flows on the Tuolumne River out to the ocean with no scientific evidence proving it would even help the fish species. If this proposal is implemented, water years like this would render New Don Pedro Reservoir empty, empty, with no water for fish at all. I fail to understand how this approach actually helps the fish.

Now the State board is threatening to do away with over 100 years of water rights history without even understanding the impacts to our reservoirs; and worse, what it would do to the people of the valley that will suffer from these short-sighted policies.

What this State needs is more storage, and we need it now, not after a few more years of study, but right now. Provisions in the House bill I authored allow for construction of millions of acre feet of new water to begin immediately. I included a provision to allow for more storage at New Melones Reservoir at no cost to the taxpayer. I also included a provision seeking to do a pilot study on the impacts predator fish have on listed species.

It is all very disappointing because while I am offering solutions to assist our State's water problems, all of my common-sense provisions have been met with either resistance and dismissal by the Senate or the President and the Governor. Instead of offering assistance to solve our problems long term, all I hear about is global warming and short-term fixes that should have already been implemented to lessen the blow in future years.

To further exacerbate an already difficult water year, I continue to hear from water managers that the last few storms that have brought needed rain to the valley has not been captured, and the

State and Federal projects continue to miss important opportunities to store more water. On March 10, the Delta Vision Foundation urged, and I will quote, "the State of California and U.S. Bureau of Reclamation to reach agreement immediately on an emergency order to allow increased pumping to capture needed water for agriculture, industry and communities." It goes on to state that, "regulators missed the opportunity during the last series of big storms to export more water when there are high levels of water flow in the bay delta ecosystem."

I look forward to hearing from our witnesses today about the very real and predictable impacts that we are facing, what the State and Federal Governments are doing to improve water supplies, and ideas for short- and long-term solutions to protect California from facing this problem next year and year after year.

I yield back.

[Applause.]

The CHAIRMAN. I thank the gentleman for his statement.

And last but certainly not least, I want to recognize the newest member of the California delegation from the valley, Mr. Valadao, who is the author of the legislation that many of us have supported that has passed that provides long-term solutions to the California water problems.

Mr. Valadao, you are recognized for 5 minutes.

STATEMENT OF THE HON. DAVID G. VALADAO, A REPRESENTATIVE IN CONGRESS FROM THE STATE OF CALIFORNIA

Mr. VALADAO. Thank you, Chairman Hastings, and thank you for taking the opportunity and the time to come out here and spend some time with our constituents here in the valley, and Congresswoman Lummis from Wyoming. I really enjoyed driving her around yesterday and pointing out some of the issues that we are facing here in the valley.

As many of you know, I am a farmer here still in the valley. I farm just south of Hanford. And in those years that I farmed, as I still do today, we have had years that we get extra rain, and we will get calls from our water district begging us to take water onto our rain-soaked and flooded fields. Yes, and it really happens, and it is something that people forget about.

When we had those really bad rains a few years back, we were getting calls, and my fields were literally flooded with water. Our crops were on the verge of dying, but they are begging us to take water. That is a situation where water infrastructure makes a difference, because that water, instead of being forced upon us, could have been held back for a year like now. And then the water that we did have in storage was wasted, slowly. You hear about releases up in McClintock's district and other parts of the State where water is being released during the winter; or there is water, just like earlier this year, that was released out into the ocean and not captured and stored.

In a year like today, when we are literally at zero percent allocation, where the well drillers in our area are tied up for at least a year-and-a-half, where some farmers are even getting to the point of buying their own drilling rigs, we have water flowing out into the ocean, water just being wasted for absolutely no good reason, because of law.

What needs to be done? A bill was passed a few years ago, H.R. 1837. A bill was passed again, H.R. 3964. Yes, everybody has been saying it is my bill. My name is at the top of the list, but I had every single Republican in the delegation sign on to that bill. Chairman Hastings sat down with all of us and said what do you need to do, how do we make you guys get something through this House again? He said as long as we work together, I am there.

We put a bill out. We made a few changes at the request of some of the Members. We introduced it. In a few weeks it was off the House Floor and it was delivered to the Senate.

How does the legislative process work now? The Senate has to act, either take up our bill or pass the bill that was introduced not too long ago by our two Senators here in California.

As far as how this group up here has handled the situation, we have reached out, we are continuing to reach out, and we are going to continue to meet with and help in any way we can to either pass our bill through the Senate or pass their bill. But at the end of the day, until we get something to the President's desk for his signature, we cannot solve this problem.

So having hearings like this helps bring more attention to what is going on in the valley to make sure there is a face with the situation that we have here. When we go back to Washington, it is a fight because we can't afford to fly all of you back there and help us bring that message. And you do see those rallies on TV, and those are helpful, and that is why people do that.

So the fact that we packed this place like we have today, standing room only, and the national attention on this issue, it helps us push this issue forward. The sad thing is, people look at this from other parts of the State. I have made a lot of friends in other States, and they always point to us and say if we don't fight, if we don't do what is right today, we are going to end up like California, and they are where our food comes from, and it is sad. It is something that—I don't want to be their bad example. I don't like watching our Members, or even parts of our State, the representatives from parts of our State, not want to help, not want to be part of the solution.

And it is a challenge, but we have a great team up here. I have had the opportunity as the new guy—and that is why I am sitting on the end——

[Laughter.]

Mr. VALADAO [continuing]. But as the new guy, to be part of the solution and to work with them and push this forward, and we are going to continue to fight. We are not giving up, and we are going to keep doing everything we can to bring something to the President's desk and get his signature.

So thank you, and thank you, and I yield back.

[Applause.]

The CHAIRMAN. I thank the gentleman.

We have a panel here of 10 people that had been invited, and I will just briefly introduce them, and then I will take a little bit more time to introduce where they are all from.

We have Mr. Mark Watte from Tulare; Ms. Sylvia Chavez from Huron; Mr. Larry Starrh from Shafter; Mr. George Delgado from Firebaugh; Mr. Tom Coleman from Madera; Mr. David Murillo from Sacramento; The Honorable Felicia Marcus from Sacramento; we have Ms. Janelle Beland from Sacramento; Mr. Steve Knell from Oakdale; and Mr. Kole Upton from Chowchilla.

Now, let me, for those of you who have not testified in front of the committee, we ask you with the invite to submit a written testimony. Your full testimony will appear in the record. However, because of time constraints, we try to keep the oral remarks within the 5-minute rule, as we call it.

Up here is a light that has a green light, a yellow light, and a red light. That is really, really significant in committee hearings, I will tell you, because what it says, when the green light is on, you are just doing swimmingly well. But when the yellow light comes on, that means you are within a minute of your 5-minute time. And then when the red light comes on—well, we just don't want to go there, OK?

[Laughter.]

The CHAIRMAN. And so what I would like all of you to do is keep your oral remarks obviously within the 5 minutes. But if your thought goes farther, we will obviously be somewhat flexible.

So I would now like to introduce Mr. Mark Watte from Tulare, California.

Mr. Watte, you are recognized for 5 minutes.

STATEMENT OF MARK WATTE, TULARE, CALIFORNIA

Mr. WATTE. Good morning, Chairman Hastings and members. My name is Mark Watte, better known now I think as the cheeseburger man.

My grandfather emigrated from Belgium in 1909, share-cropping in southern California for 50 years. After World War II the ranch was developed into housing, and my father and uncle moved to Tulare County in 1958. They started with 560 acres, successfully farmed and split their partnership in 1984. My brother and I bought the business from our father in 1986, and we have grown significantly since then. Together with my brother Brian, nephew Matthew, and son-in-law Jason, we milk 1,000 cows, raising 18,000 calves, and we farm 4,500 acres, which is about 7 square miles, of diversified row crops and have recently started planting pistachio trees. The downside to that growth is we are like Norm and Cliff at Farm Credit, they know our name.

For the last two decades, as a result of an onslaught of overreaching rules and regulations spurred on by environmental activists, we have lost and continue to lose huge amounts of our potential surface supply for, in many cases, no tangible results. Overall, the fish populations are no better off and perhaps worse. The activist answer is just to flush more water. Here are a couple of examples of what I am talking about.

Hundreds of thousands of acre feet of water flows to the Sacramento Delta to improve water quality, quality degraded by neighboring cities dumping low-quality sewage into the river. Solution by dilution is not an answer. By the way, the water coming from Hetchy-Hetchy reservoir that supplies San Francisco, home of Senators Boxer and Feinstein and Congresswoman Pelosi, does not contribute any water toward this effort, and today that reservoir has one of the highest percent of capacity in the whole State.

Pumping water through the delta is one of the key components of our statewide system. These pumps are severely restricted ostensibly to protect a 3-inch bait fish that isn't even indigenous to the delta. This is only a ruse used by these same activists. They don't care about fish. They just don't want us to get our entitled water. If any of these groups really cared about fish, they would be talking about a huge stressor on salmon and smelt population, the striped bass.

Another big chunk of eastside water is now being lost to support a river restoration effort that after several years is failing miserably. I believe another presenter will be discussing this in more detail.

In 1992, CVPIA committed 1.2 million acre feet a year for environmental uses that anyone today would be hard pressed to show any tangible results. Also, none of that water used for any of these environmental programs were paid for or held accountable for any benefits.

These are but a few of the leaks on our developed water supply in California. These, coupled with a 3-year drought, has brought us to where we are today. President Obama, Governor Brown, and Senator Feinstein have put forward initiatives to spend hundreds of millions of dollars to mitigate drought damage. We don't need money; we need water. It is preposterous to offer billions of dollars to combat climate change, global warming, whatever name you want to put on it, and think that will help the California water supply.

Any meaningful, substantive progress in improving our situation has to begin with some common sense injected into the entire endangered species discussion. We need to look at where ESA has worked. There are many cases of this. But where is the law being used for reasons other than species recovery? Without some reasoned middle ground in the debate, no real progress will be achieved.

Another significant aspect relative to the overreach of the ESA is the huge increased cost of building any water-related projects. We have totally lost our sense of balance between making significant positive advances with minimal effects.

Case in point. I am Thomas Edison. I just invented electricity. I am now filing my EIR. Part of it is going to read something like this. I have invented an energy that will revolutionize the way we live. But to transmit this energy, we will need to build transmission lines along our roads. To do this, we will have to cut down a tree, make a pole, and they will be about every 300 feet. The bottoms of the poles are going to need to be treated with something so they don't rot and fall over in a couple of years. Distracted drivers could run off the road and cause themselves a lot of harm. The overhead wires will not be very attractive, and once in a while an endangered Swainsons hawk will touch two of the wires and we will have a barbecued hawk.

Would we have electricity today? And if we did, at what cost? Ask yourself if having power is worth it. Of course it is. So is having an abundant, affordable, and safe food supply.

In conclusion, Congressman Nunes and I were recently featured in a far-reaching article in *The Wall Street Journal* that pointed out many of the absurdities of farming in California. The response to the article has been overwhelmingly positive. This country still has many commonsense people. It makes me hopeful that perhaps there is a realization that the pendulum of extremism needs to be moderated.

Thank you, Mr. Chairman and all of the Members, for taking time to listen to our concerns. And I would like to submit the mentioned *Wall Street Journal* article into the record.

[The prepared statement of Mr. Watte follows:]

PREPARED STATEMENT OF MARK WATTE, TULARE, CALIFORNIA

Good Morning Chairman Hastings and Members.

My grandfather emigrated from Belgium in 1909, share cropping in southern California for 50 years. After World War II it was developed into housing and my father and uncle moved to Tulare County in 1958. They started with 560 acres and eventually split their partnership in 1984. My brother and I bought the business from our father in 1986 and have grown significantly since then. Today with my brother Brian, nephew Matthew, and Son-in-Law Jason; we milk 1,000 cows, raise 18,000 calves and farm 4,500 acres (7 square miles) of diversified row crops and have more recently started planting pistachios. We are like Norm and Cliff at our Farm Credit office, they all know our name.

I currently serve on seven boards and commissions, of which five are directly related to water. Married for 41 years, 3 married daughters and 10 nearby grandchildren.

I don't know of anyone that is more committed or passionate about our area and way of life than myself.

For the last two decades, as a result of an onslaught of over-reaching rules and regulations spurred on by environmental activists, we have lost and continue to lose huge amounts of our potential surface water supply, for, in many cases, no tangible results. Overall, the fish populations are no better off and perhaps worse. The activist answer is to flush more water. Here are a couple of examples of what I am talking about.

1. Hundreds of thousands of acre-feet of water flows to the Sacramento Delta to improve water quality-quality degraded by the neighboring cities dumping low quality sewage into the river. Solution by dilution is not an answer. By the way, the water coming from Hetchy-Hetchy reservoir that supplies San Francisco, home of Senators Boxer and Feinstein and Congresswomen Pelosi, does not contribute any water toward this effort and today that reservoir has one of the highest percent of capacity in the entire State.
2. Pumping water through the delta is one of the key components of our State wide water system. These pumps are severely restricted ostensibly to protect a 3 inch bait fish that isn't even indigenous to the delta! This is only a ruse used by the same activists. They don't care about fish, they just don't want us to get our entitled water. If any of these groups really cared about fish they would be talking about a huge stressor on salmon and smelt population, the striped bass.
3. Another big chunk of east side water is now being lost to support a river restoration effort that after several years is failing miserably. I believe another presenter will be discussing this in more detail.
4. In 1992 CVPIA committed 1.2 million acre feet to environmental uses that anyone today would be hard pressed to show any tangible results. Also, none of the water used for any of these ''environmental programs'' were paid for or held accountable for the benefits achieved.

These are but a few ''leaks'' on our developed water supply in California. These coupled with a 3-year drought has brought us to where we are today. President Obama, Governor Brown, and Senator Feinstein have put forward initiatives to spend hundreds of millions of dollars to mitigate drought damage. WE DON'T NEED MONEY-WE NEED WATER! It is preposterous to offer to billions of dollars to combat climate change/global warming and think that will help the California water supply.

Any meaningful substantive progress in improving our situation has to begin with some common sense injected into the entire endangered species discussion.

What we need to look at is what has worked—there are many cases of this—but where is the law being used for reasons other than species recovery. Without some reasoned middle ground in the debate no real progress can be achieved. Another significant aspect relative to the overreach of the ESA is the huge increased cost of building any water related projects. We have totally lost our sense of balance between making significant positive advances with minimal negative effects.

Case in point. I'm Thomas Edison and just invented electricity. I am now filing my EIR, which will include among many others State and Federal fish and game, NEPA, CEPA. It would read like this—I have invented an energy source that will revolutionize the way we live. But to transmit this energy we will need to build a transmission line along our roads. To do this we will need to cut down a tree to make a pole and they will be every 300 feet along our roads. The bottoms of the poles will need to be treated so that they will not rot; distracted drivers could run off the road and kill themselves running into a pole. The overhead wire will not be attractive and once in awhile an endangered Swainsons Hawk will touch two of the wires and we will have a BBQ'd hawk. Would we have electricity today, and at what cost? Ask yourself if having power is worth it? Of course it is. So is having an abundant affordable food supply.

In conclusion, Congressman Nunes and I were recently featured in a far-reaching article that pointed out many of the absurdities of farming in California. The response has been overwhelming positive. This country still has many common sense people. It makes me hopeful that perhaps there is a realization that the pendulum of extremism needs to be moderated.

Thank you Mr. Chairman and all of the members for taking time to listen to our concerns.

The CHAIRMAN. Without objection, it will be part of the record.

Mr. WATTE. Thank you.

The CHAIRMAN. Thank you very much, Mr. Watte.

[Applause.]

The CHAIRMAN. Next I will recognize Ms. Sylvia Chavez, who is the Mayor of the city of Huron.

Mayor Chavez, you are recognized for 5 minutes.

STATEMENT OF THE HON. SYLVIA V. CHAVEZ, MAYOR, CITY OF HURON, CALIFORNIA

Mayor CHAVEZ. Mr. Chairman and Members of Congress, good morning and welcome to the Central Valley. Thank you for coming today to learn about the impacts the drought is having on our communities.

The CHAIRMAN. Speak more closely into the mic. There you go.

Mayor CHAVEZ. I hope when you leave today you will take a better understanding that for our communities in the Central Valley water is jobs and water is life.

My name is Sylvia V. Chavez and I am the child of farm workers, as were my parents before me. As a child, I worked in the fields during my summer vacations to earn extra income. My mother and father worked in the fields out of necessity to provide a better life for our family. Their hard work showed our eight siblings and me what work ethic is truly all about.

Today, I am the Mayor of the city of Huron. Huron has a population of just under 7,000 people and is located 60 miles southwest of Fresno. Our population is 97 percent Hispanic. The majority of our residents are connected to agribusiness either directly or indirectly. Because our city is so far away from Fresno and the other population centers, many times we are forgotten. Yet, like many other small valley towns, when it comes to putting food on the dinner table, it is communities like ours that fill our Nation's stomachs with many of the everyday foods Americans take for granted.

You see, my city, like many valley towns, is surrounded by agriculture. Local farmers plant, irrigate, and harvest their crops with the help of Huron's residents. Then, the people of Huron pack and transport valley commodities to market. If you did not grow up in the valley or have not traveled here before, you may be unaware that the lettuce and tomato in your garden salad or the toppings on your McDonald's hamburger come from here in the San Joaquin Valley.

Our region was blessed with fertile farm land. What we grow is not simply transported to other parts of our Nation. Our commodities are shipped across the globe. The next time you put sauce on your spaghetti, remember that 95 percent of the processing tomatoes in the United States are grown in Huron. Let's face it, in Huron, we feed the world.

As much as our community is tied to agriculture, we are equally tied to water. In 2009, when water allocations reached as low as 10 percent, Huron's unemployment rate climbed to almost 40 percent. Businesses who normally hired as many as 3,500 farm workers in previous years needed less than 600 because of the drought. As a result of the 2009 drought, many in my community were forced into food lines just to feed their families. The drought we face today is by far more serious. In fact, the drought we face today has put my community's ability to turn on its faucets in jeopardy.

In the city of Huron, we purchase our water from the Bureau of Reclamation. This year we were notified that Huron will receive an allotment of 649 acre feet for fiscal year 2014–2015. For my city, whose historical usage is 1,125 acre feet per year, this year's allotment represents a shortage of 476 acre feet. Because of record drought conditions, my city is already tapping into its water allotment. Our local water managers have become concerned enough that the Huron City Council recently passed a resolution restricting water use on residential and industrial properties. Our community truly understands the value of water, and the council is confident the city's residents will conserve all the water they can, but will it be enough?

Today, I am calling on you, the Members of Congress gathered here today, to provide a solution to the man-made drought that is crippling my community. If the drought is not dealt with quickly and appropriately and actions are not taken to better balance the needs of our community, and communities like it, with the needs of the delta fish, the inaction will truly threaten my community's existence. I fear continued drought and water diversions will make our agricultural community a thing of the past. Our residents will be without jobs and incomes, and our city will suffer the consequences. In short, our economy will fail.

Solving the water crisis is so urgent to my city that when I told my friends and neighbors that I was coming here today to testify, many in my community wanted to share their stories as well. They wanted to tell you, their elected representatives in Congress, how the water crisis is impacting them. Today I have brought with me letters from many of my neighbors who wanted to have their voices heard. I hope you will take these letters with you, listen to their stories too, and use the knowledge they share to inform your decisionmaking in Washington.

For many, the unemployment in the town of Huron may be forgotten once we leave here today, but it shouldn't be. For my community, water is about jobs and the opportunity to thrive. But what about you and your communities? To that we say, what about your dinner table? Congress must act soon to provide drought relief. Remember, it is communities like Huron, California that feed the world.

In ending, I wanted to state something that my granddaughter told me last night. As I was speaking with my husband about this——

The CHAIRMAN. Quickly, quickly. Go ahead, please.

Mayor CHAVEZ. As I was speaking to my husband about this, my granddaughter stated, "Grandmother, you need to make them hear. You need to make them understand. My friends are afraid. Their parents have been talking about losing their jobs because of no water, no jobs for them, and they are thinking of moving." She was very serious, and this is coming from my 10-year-old granddaughter.

She understands. She is 10 years old, and she understands the impact this is having on our community and her friends and their families and how to provide food on the table. And she said, "Papa and you and mama put food on our table, but our friends are worried about how they are going to be fed this summer."

Thank you.

[The prepared statement of Mayor Chavez follows:]

PREPARED STATEMENT OF THE HONORABLE SYLVIA V. CHAVEZ, MAYOR, CITY OF HURON, CALIFORNIA

Mr. Chairman and Members of Congress, good morning and welcome to the Central Valley. Thank you for coming here today to learn about the impacts the drought is having on our communities. I hope that as you leave today you will take with you a better understanding that for many communities in the Central Valley water is jobs and water is life.

My name is Sylvia V. Chavez and I am the child of farm workers, as were my parents before me. As a child, I worked in the fields during my summer vacation from school to earn extra income. My mother and father worked in the fields out of necessity and survival to provide a better life for our family. Their hard work showed my eight siblings and I what work ethic is truly all about.

Today, I am the Mayor of the city of Huron, California. Huron has a population of just under 7,000 people and is located 60 miles southwest of Fresno. Our population is 97 percent Hispanic. The majority of our residents are connected to agribusiness either directly or indirectly. Because our city is so far away from Fresno and the other population centers, many times we are forgotten. Yet, like many other small valley towns, when it comes to putting food on the dinner table, it is communities like ours that fill our Nation's stomachs with many of the everyday foods Americans take for granted.

You see, my city, like many valley towns, is surrounded by agriculture. Local farmers plant, irrigate, and harvest their crops with the help of Huron's residents. Then, the people of Huron pack and transport valley commodities to market. If you did not grow up in the valley or have not traveled here before, you may be unaware that the lettuce and tomato in your garden salad or the toppings on your McDonalds hamburger burger come from right here in the San Joaquin Valley. Our region has been blessed with fertile farm land. What we grow is not simply transported to other parts of our Nation. Our commodities are shipped across the globe. The next time you put sauce on your spaghetti, remember that 95 percent of the processing tomatoes in the United States are grown in Huron. Let's face it, in Huron,

WE FEED THE WORLD!

As much as my community is tied to agriculture, we are equally tied to water. In 2009, when water allocations reached as low as 10 percent, Huron's unemployment rate climbed to almost 40 percent. Businesses who normally hired as many as 3,500 farm workers in previous years needed less than 600 because of the drought. As a result of the 2009 drought, many in my community were forced into food lines just to feed their families. The drought we face today is by far more serious.

In fact, the drought we face today has put my community's ability to turn on its faucets in jeopardy. In the city of Huron, we purchase our water from the Bureau of Reclamation. This year we were notified that Huron will receive an allotment of only 649 acre feet for fiscal year 2014–2015. For my city, whose historical usage is 1,125 acre feet per year, this year's allotment represents a shortage of 476 acre feet. Because of record drought conditions, my city is already tapping into its water allotment. Our local water managers have become concerned enough that the Huron City Council recently passed a resolution restricting water use on residential and industrial properties. Our community truly understands the value of water and the Council is confident the city's residents will conserve all the water they can, but will it be enough?

Today, I'm calling on you, the Members of Congress gathered here today, to provide us a solution to the manmade drought that is crippling my community. If the drought is not dealt with quickly and appropriately and actions are not taken to better balance the needs of my community, and communities like it, with the needs of delta fish, the inaction will truly threaten my community's existence. I fear continued drought and water diversions will make our agricultural community a thing of the past—our residents will be without jobs and incomes and our city will suffer the consequences. In short, our economy will collapse.

Solving the water crisis is so urgent to my city that when I told my friends and neighbors I was coming here today to testify many in my community wanted to share their stories as well. They wanted to tell you, their elected representatives in Congress, how the water crisis is impacting them. Today I have brought with me letters from many of my neighbors who wanted to have their voices heard. I hope you will take these letters with you, listen to their stories too, and use the knowledge they share to inform your decisionmaking in Washington.

For many, the unemployment in the town of Huron may be forgotten once we leave here today, but it shouldn't be. For my community water is about jobs and the opportunity to thrive. But, what about you and your communities? To that I say, ''what about your dinner table?'' Congress must act soon to provide drought relief. Remember, it is communities like Huron, California that FEED THE WORLD!

The CHAIRMAN. Ms. Chavez, thank you very much.

You referenced some letters. Would you like to have those letters part of the record?

Mayor CHAVEZ. Yes, please. I gave them to Congressman Valadao's office.

The CHAIRMAN. OK. I will recognize Mr. Valadao.

Mr. VALADAO. Mr. Chairman, I ask unanimous consent that the letters Mayor Chavez brought with her today and the letters of my constituents on the impacts of the current drought be entered into the record.

The CHAIRMAN. Without objection, they will be part of the record.

Mr. VALADAO. Thank you.

The CHAIRMAN. Now I recognize Mr. Larry Starrh, who is co-owner of Starrh——

[Applause.]

The CHAIRMAN. Mr. Larry Starrh, who is co-owner of Starrh and Starrh Farms in Shafter, California.

Mr. Starrh, you are recognized for 5 minutes.

STATEMENT OF LARRY STARRH, CO-OWNER, STARRH AND STARRH FARMS, SHAFTER, CALIFORNIA

Mr. STARRH. Thank you, Mr. Chairman, and thank you, Members of the House.

I don't know that I should say anything. I mean, you all got it. Everything that you said here at the dais needs to happen. So you know what to do, and you are doing it, so I don't know if what I can add is going to be much help, but I am going to read it anyway because I wrote it.

[Laughter.]

Mr. STARRH. But seriously, you know what needs to happen. We need the water, and you know how to get it done. So I wish you well, and get it done.

My name is Larry Starrh, and I am a partner in Starrh Family Farms. Our farm is located in Kern County, and my partners include my father Fred Starrh, my brother Fred Starrh II, and my brother-in-law Jay Kroeker. Our office and all our bookkeeping is run by my two sisters, and most recently my niece. My oldest son is now farming with us and is excited by the opportunity for a future in farming. Good luck. My youngest son is also looking forward to being a farmer and has said so since he was a toddler. God bless him, right?

We employ 46 full-time employees, with an average time of service to our ranch of over 20 years, good men who have dedicated years of their life to provide for their families and to help make our farm successful, and we are grateful to them. I tell you this so you can get a glimpse of who and what I represent. I am not alone when I sit here.

We farm close to 9,000 acres, primarily almonds and pistachios. The bulk of our ranch is located in the Belridge Water Storage District, which lies on the west side of Kern County, and it is completely reliant on the State Water Project for providing water.

I know this hearing is about water, and my family has been farming for over 80 years, and according to what I have been told by the experts, this is the driest year on record in California. Due to the lack of water this year, my family and I had to make the hard decision to dry up and let die close to a thousand acres of producing almond trees. As well, we will continue to keep fallow another 2,000 acres of open ground, ground that we have had to keep idle for close to 8 years because of water shortages, shortages that were created and sustained by regulations, regulations that have been imposed and brandished like weapons, weapons that are built on myths and hyperbole.

In the last 12 years, our farm hasn't received 100 percent allocation of entitlement water once. But every year we have to pay for 100 percent of that water even though we don't receive it. Every year the State Water Project takes water out of the system for environmental needs. Every year the State bills us for that water, and we have to pay it. The people who have to pay it are the water users it is taken from, and the government recognizes that word. It is a taking, and it is OK. I don't get it.

As a grower, the challenges get even greater. Reliability of contracted water is non-existent. You can't make crop decisions. You can't make labor decisions. You can't make them until the last minute, or ever. And on top of that, we have to try to source dry-year water, if it is even available, to buy at who knows what the cost will be. And the cost can double or triple. This year we paid $1,250 an acre foot for water, and that was at a bid price. We had to bid for that water. And until we did it, we said we have to do it because we have to keep our trees alive, the rest of our trees alive.

We can't sustain this way of doing business. The water system in California is crippled and needs to be repaired.

Two weeks ago at our water district meeting, it was reported that the Sacramento River was running at flood stage, flood stage, right? But we couldn't move the water because the San Joaquin River was low. I don't get it. I really don't get it.

Last year, 800,000 acre feet of State Water Project water was released to the ocean instead of being stored. Last year we bought dry-year water, and we purchased it, and it was almost stranded in Oroville because of mismanagement. Had it rained this year as normal, we would have lost it.

The water system in California is crippled and needs to be repaired.

In my naive world, water is life. Water creates life. Water sustains life. Sadly, in the real world, I think water is about power. Water is a weapon. Water is a hostage. Our water system is battered and broken and has been hijacked by the unreasonable, and we need help.

This year we are in a drought caused by nature. I know that. But in the years prior to this, the droughts we have suffered were imposed on us by the illogical and the senseless. And I know that you folks understand this, and I know that I am just reiterating things, but we do need to look at this seriously. Three years ago I sat in the audience and I listened to a lot of these same things, and I thought, wow, we are on the road, we are going to fix this. Here we are again. We are on the same road.

I know you understand this, so I don't know how you get any further. I thank the Lord that you are doing it, and I appreciate your time here today, and I appreciate you allowing me to testify in front of you.

Thank you, Mr. Chairman. That is all I have.

[The prepared statement of Mr. Starrh follows:]

PREPARED STATEMENT OF LARRY STARRH, CO-OWNER, STARRH AND STARRH FARMS, SHAFTER, CALIFORNIA

My name is Larry Starrh, I am a partner in Starrh Family Farms. Our farm is located in Kern County and my partners include my father Fred Starrh, my brother Fred Starrh II, and my brother in-law Jay Kroeker. Our office and all our book keeping is run by my sisters and recently my niece and my oldest son is now farming with us and is excited by the opportunity for a future in farming. My youngest son is also looking forward to being a farmer and has said so since he was a toddler. We employ 46 full time employees with an average time of service of over 20 years. Good men who have dedicated years of their life to help make our farm successful and we are grateful to them. We farm close to 9,000 acres primarily almonds and pistachios. The bulk of our ranch is located in the Belridge Water Storage District and is completely reliant on the State Water Project for providing water.

This meeting is about water. My family has been farming for over 80 years, and according to what I have been told by the experts this is the driest year on record in California! . . . Due to lack of water this year my family and I had to make the decision to ''dry up and let die'' close to a thousand acres of producing almond trees, as well as keeping fallow another two thousand acres of open ground. Ground that we have had to keep idle for close to 8 years because of water shortages. Shortages that were created and controlled by regulations, regulations that have been imposed and brandished like weapons! On the State Water Project water has been "taken" out of the system to protect environmental needs, to add insult to injury the bill for that water is paid for by the people who the water was taken from. Our farm hasn't received a hundred percent allocation for water in 18 years but we have been charged and have had to pay for 100 percent every year. We pay for water we don't get. Then we have to try and source "dry year water" to buy which can cost double or triple, or like this year 10 times more than the base price. We can't sustain this way of doing business, the water system in California is crippled and needs to be repaired.

In my world water is life. Water creates life, water sustains life. Sadly in the real world water is about power, water is a weapon, water is a hostage. Our water system is battered and broken and has been kidnapped/hijacked by the unreasonable, and we need help! This year we are in a drought caused by nature, I know that, but in the years prior to this the droughts we have suffered were imposed on us by the illogical and senseless. Thank you! Thank you for your commitment and your understanding that water is life! And thank you for trying to find a reasonable solution!

The CHAIRMAN. Thank you, Mr. Starrh.

[Applause.]

The CHAIRMAN. Thank you, Mr. Starrh.

I will now recognize Mr. George Delgado, owner of Delgado Farming in Firebaugh.

Mr. Delgado, you are recognized for 5 minutes.

STATEMENT OF GEORGE DELGADO, DELGADO FARMING, FIREBAUGH, CALIFORNIA

Mr. DELGADO. Delgado.

[Laughter.]

The CHAIRMAN. I knew I would mess that up.

Mr. DELGADO. Chairman Hastings and members of the committee, I thank you for the opportunity to come here to testify before you today on one of the most important issues facing my community.

My name is George Delgado. I am a farmer on the west side of the Central Valley in Firebaugh, a small community in western Fresno County where I have lived my entire life. My experiences in agriculture on the west side began long before I started farming. As a young man, I learned to work in the fields, whether it was chopping weeds, picking cotton or tomatoes by hand. My father told me that if I didn't want to work in the fields the rest of my life, I should get an education, so I did. I attended Fresno State. I earned a degree in agricultural science.

I continued to work weekends and summers for west side farmers who gave me an opportunity to work so that I could pay for my college education.

In 1978, I leased some land near Firebaugh and began farming on my own. A few years later, I leased an additional 300 acres on the historic Sam Hamburg Ranch, where I continue to farm to this day. Presently, I own and farm several hundred acres of almonds, cherries and cantaloupe in the San Luis Water District, Pacheco Water District, and Westlands Water District. Each of these districts receives water from the Central Valley Project and has been greatly affected by the drought and the environmental restrictions.

The heart of the Central Valley Project is the Sacramento San Joaquin River Delta. Water naturally flows to the delta from reservoirs in northern California, where it is pumped into both the Central Valley and State Water Project man-made canals and aqueducts. Unfortunately, water conveyance through the delta has presented significant challenges to State water systems.

Besides water quality in the delta, the environmental laws and continuous litigation brought largely by environmental special interest groups have constrained California's water system. The Federal Endangered Species Act has been the major environmental driver in water supply litigation. Efforts to protect species such as the delta smelt have created a tremendous amount of uncertainty in our annual water supply. Over the last decade, millions of acre feet of water have been diverted away from human use to save these species. Environmentalists have repeatedly blamed the operations of the delta pump for causing the delta population decline. Yet, they have ignored other proven factors, including predation by non-native fish such as the striped bass and the discharges of toxic sewage into the delta from the cities.

The pumping has been stopped even in wet years to protect fish, yet the delta ecosystem continues to be in decline. Water that could have been stored for use in dry years such as the current year has been lost forever. Unfortunately, protection of the delta smelt is not the only reason water has been taken away from Central Valley Project water users. When I began farming, west side farmers could expect to receive 100 percent of their contracted water supply each year, except in the years of the most extreme drought conditions.

However, since the passage of the Central Valley Project Improvement Act of 1992, over 1 million acre feet of water each year has been taken away from irrigated agriculture and dedicated to fish and wildlife uses. As a result, in an average water year, most farmers on the west side receive less than half of their contracted water allocation.

The CVPIA has had a devastating effect on our communities, especially in years of below-average rainfall. Hundreds of thousands of acres have been fallowed. Unemployment and crime rates have dramatically increased. Here in the San Joaquin Valley, water equals jobs, not just farm jobs but off-farm jobs. It is sad to see, here in the Nation's food basket, so many people forced into food lines to receive food that is likely grown in China or other parts of the world.

As farmers, a natural occurring drought is an acceptable risk of our chosen profession. But a drought caused by restrictive legislation is very difficult to understand. The CVPIA was enacted while California was experiencing the effects of a long-term drought, and it was intended to encourage water conservation, increase the use of water transfers, and to provide additional water for fish and wildlife. However, it has amounted to little more than legally stealing water from the farmers to dilute discharges of sewage, metals and chemicals dumped into our rivers, the delta, the San Francisco Bay; and on the west side, we have invested millions of dollars in installing state-of-the-art irrigation systems to improve water quality and conserve our diminishing water supply, yet we are blamed for the continued decline of the delta and our waterways, not the polluters who refuse to live by the laws that they impose on us.

Much can be done to improve our situation here in the Central Valley. The Endangered Species Act must be reformed to strike a reasonable balance between people and fish. The CVPIA must be amended to encourage balance between the needs of our cities, farmers, and the environment.

I thank you for the opportunity to share my story with you today. Only a united Congress and a President can make the necessary changes and enact legislation to give us short- and long-term solutions to our water issues. Please take what you learn here today back to Washington and, working together, use it to help provide the relief that the valley needs.

[Applause.]

[The prepared statement of Mr. Delgado follows:]

PREPARED STATEMENT OF GEORGE DELGADO, DELGADO FARMING, FIREBAUGH, CALIFORNIA

Chairman Hastings and members of the committee, I thank you for the opportunity to come here to testify before you today on one of the most important issues facing my community.

My name is George Delgado. I have lived on the west side of the Central Valley all my life in Firebaugh, western Fresno County. I attended local schools in Firebaugh, California and completed my education at Fresno State University earning a degree in Agricultural Science.

My experiences in agriculture on the west side go back to well before I started my first farm. As a young man, I picked cotton and tomatoes and chopped weeds by hand. Working summers and on weekends in the fields for west side farmers gave me the opportunity to work as I attended school and helped me earn the money I needed to complete my education. My career as a farmer started in 1978 when I leased my first field near Firebaugh. Later, I leased an additional 300 acres on the historic Sam Hamburg Ranch, first cultivated in 1936. Presently, I own and farm almonds, cherries, and cantaloupes in Westlands Water District, San Luis Water District and Pacheco Water District, all of which receive their water from the Federal Central Valley Project.

The hub of California's Central Valley Project is the Sacramento-San Joaquin River Delta. Here, water from reservoirs in the northern portions of the Central Valley and State water projects is conveyed through natural channels to pumps that feed the man-made canals and aqueducts that carry water to the west side and down toward the southern portions of our State. Unfortunately, using the delta's natural channels to convey water through the system has shown itself to be the equivalent of using an unimproved dirt road as an interchange on our Federal interstate system and it has imposed significant challenges on the State's water systems.

Environmental statutes and litigation, brought largely by environmental special interests, have led to serious water conflicts in California. The Federal Endangered Species Act [ESA] has been the major environmental driver in water supply litigation. Of the over 1,300 species listed under the Endangered Species Act in the United States, over 300 are in the State of California. During the past 10 years, trillions of gallons of water have been diverted away from human use to environmental purposes to "save" these species.

Recent litigation on protecting delta smelt, a 3-inch fish native to California's Sacramento-San Joaquin River Delta, has taken hundreds of thousands of acre feet away from our communities each year. Environmentalists have consistently blamed the delta pumps as the cause for smelt population decline. Yet, they continue to ignore numerous other factors, including predation by nonnative fish such as the Striped Bass and the discharge of toxic sewage into the delta, all of which have been shown to contribute significantly to smelt decline. Unfortunately, the delta smelt is not the only reason water has continued to be taken from the valley.

When I started farming on the west side, farmers could expect to receive 100 percent of their contracted water supplies year-in and year-out, except in years of the most extreme drought conditions. However, since the passage of the Central Valley Project Improvement Act of 1992, more than 1.2 million-acre feet of water annually—enough to irrigate over 340,000 acres of farmland—have been redirected away from irrigation to fish and wildlife uses. As a result, in an average water year, most farmers on the west side expect to receive less than 40 percent of their allocation from year to year.

Agriculture in my part of the valley has been devastated by the Central Valley Project Improvement Act. As farmers, we can accept natural droughts as a risk of our chosen profession but a drought caused by legislation that takes away our water is very difficult to understand. The Central Valley Project Improvement Act was enacted while California was experiencing the effects of a long-term drought and many of the provisions in the act were aimed at conserving water, increasing the use of water transfers, and providing additional water for fish and wildlife purposes.

As someone who makes his living off the land, I am all too aware of the need for all of us to be good stewards of the earth. However, part of being a good steward is ensuring that scarce resources are allocated in the most efficient and effective means possible, striking careful balances. That means ensuring there is enough water available for both fish and families. The continued decline of threatened and endangered species in the State in the face of CVPIA's water reallocation has led me, and many others who make their livings on the west side, to ask whether taking away our water for fish and wildlife has had a meaningful impact on our environment.

Growing up, I was taught that the purpose of our Government is to help farmworkers, those in agriculture related professions, and farmers as they struggle to grow America's vegetables, fruits, nuts and other food products. Those of us who farm in the valley are proud to say we feed the world. However, the continued manmade drought has left many families in communities up and down the valley unable to feed themselves.

Here in the San Joaquin Valley, water equals jobs. In 2009 during the last water crisis, hundreds of thousands of acres were fallowed, leaving many thousands unemployed. Our communities saw unemployment rates reach well over 40 percent and crime rates hit record highs. Here, in the Nation's food basket, many of our friends and neighbors were forced into food lines to receive Chinese produce.

Like many of my friends and neighbors, I am afraid the CVPIA is doing little more than legally stealing water from farmers. Here in the Central Valley, we work every day to conserve every drop of water that is delivered to us and protect our precious and quickly diminishing ground water resources so we can continue to feed the world. Although we are on the cutting edge of irrigation technology and we feed the world with the minimum water necessary, each year more water is taken from us to help clean up sewage, metals and chemicals dumped into the Sacramento River, San Joaquin River, the delta, and San Francisco Bay by polluters who refuse to keep up with the times.

Much can be done to improve our situation here in the Central Valley. California's farmers cannot continue to give up their water for "environmental purposes". The Endangered Species Act must be reformed to strike a reasonable balance that puts families first and the CVPIA has to be amended to help bring a compromise between the needs of wildlife, cities and food producers.

Thank you for the opportunity to come here today and share my story. Only a united Congress and President can work together to make these changes and enact legislation necessary to give short and long term drought relief to our communities. I hope you will take what you learn here today back to Washington and, working together, use it to help provide the relief our valley needs.

The CHAIRMAN. Thank you, Mr. Delgado. I was trying to put an "R" in your name by way of introduction, and I don't know why. I didn't see an "R", but I tried to put an "R" in there, so thank you for correcting me.

Now I want to recognize Mr. Tom Coleman. Mr. Coleman is the President of the Madera County Farm Bureau from Madera.

Mr. Coleman, you are recognized for 5 minutes.

STATEMENT OF TOM COLEMAN, PRESIDENT, MADERA COUNTY FARM BUREAU, MADERA, CALIFORNIA

Mr. COLEMAN. Thank you. Mr. Chairman and members of the committee, thank you for providing me the opportunity to testify today on the subject that is of great importance to so many. I appear today not just representing myself but thousands of people who rely on agriculture to sustain their livelihoods. It is not just growers and farmers who are affected by this situation but their employees, bankers, businessmen, and the entire community.

We have heard about the horrific effects the drought has caused in the Central Valley, and I am sure we will continue to see far-reaching consequences into the future. It is important that these effects not be minimized. But as President of the Madera County Farm Bureau, I would like to focus on one solution that has the potential to solve so many of these problems, the construction of a reservoir in the Upper San Joaquin River called Temperance Flat.

Temperance Flat is a project that has been on the books since the 1950s, first authorized by Congress in 2003 and in 2004, now only to be introduced in several pieces of legislation that have gone before or are pending in front of this committee. This storage project is unique as it is designed to be a flood storage facility only and would not impede San Joaquin River flows during normal rain years. It is also exceptional in that this project does not touch the delta directly, which makes it a prime candidate for less controversy.

Temperance Flat accomplishes multiple objectives, all of which will bring major relief to the problems described today in the following ways: increased water supply reliability and system operational flexibility for agriculture, MNI, and environmental purposes, regardless of who gets the lion's share of the water; enhanced environmental benefits through better temperature flow conditions along the San Joaquin River. The water from Temperance Flat can flow north or south as the conveyance facilities are in place already to do so. If ecosystem restoration ultimately remains impossible due to the current arrangement on the San Joaquin River, the construction of Temperance Flat Reservoir will provide a major relief for all of the system.

Finally, the local cost share associated with Temperance Flat will be generous, if not the highest available. Farmers want this project and are willing to pay for it. We don't necessarily need the Federal Government or the irrigation districts to bear the cost.

Madera County Farm Bureau's main objective is to protect its vast membership throughout Madera County. Even though all of these people's water needs are different, everyone would also benefit from having more water in the system regardless of who gets it.

The Farm Bureau appreciates the community's efforts to highlight this important issue in Congress, but we would appreciate your continued assistance and dedication toward providing a major, sweeping solution to this crisis so that it never has to happen again. Thank you.

[Applause.]

[The prepared statement of Mr. Coleman follows:]

Prepared Statement of Tom Coleman, President, Madera County Farm Bureau, Madera, California

The Madera County Farm Bureau [MCFB] is a representative member body composed of 1,200 members, 550 agricultural operations, and 170 agri-businesses. Madera County's top agricultural commodities include almonds, grapes, milk, pistachios, and cattle livestock operations. The 2013 gross agricultural value of Madera County agricultural commodities was $2,739,411,000.00—ranking the county as the

10th largest agricultural producing county in the State of California, and the 16th largest agricultural commodity sector in the world.[1] Madera County has an agricultural production acreage exceeding 2 million acres; 1.5 million of those acres belong to irrigable agricultural practices.

Historically, Madera County agricultural production has been rooted in arid rangeland grazing to the east, along with permanent crops throughout the Central Valley floor, including vines and orchards. Due to rising crop values of permanent crops since 2003 however, Madera County is now largely dedicated toward permanent crop production, including almonds, pistachios, and grapes as of 2014.[2] This transition to a high percentage of permanent crops—in some places triple plantings taking place, has occurred at an extremely rapid rate, increasing in the County's irrigation demands.

Water usage for this shift in planting activities has been significant in contributing toward the need for a conjunctive use basin; the use of groundwater as well as surface water, and has nearly doubled the amount of surface water required for irrigation of these permanent crops and tripled the amount of groundwater required to sustain the deep root bases these commodities have. A significant amount of farmed areas in Madera County are entirely dependent on groundwater—to which is in a serious overdraft condition. It is estimated that by 2017, Madera County groundwater will be overdrafted by 200,000 acre feet (AF).[3]

Agricultural conversion—land being taken out of production and dedicated toward residential or municipal purposes, has also not only slowed, but by 2013 had been reversed in Madera County. Land that was zoned for residential housing purposes in the Madera County General Plan has now been placed back into Agricultural Zoning.[4]

This new water burden associated with these agricultural practices in creating the critical groundwater overdraft condition is called subsidence in the most extreme cases. In the case of Madera County, this phenomenon occurs when so much groundwater has been pumped out that the physical sea level of the land is dropped. The upper aquifers that wells typically rely on have been depleted and growers are therefore drilling deeper—sometimes as much as 500 feet, to locate water. At this level, there is significant disruption to the Corcoran Clay layer, ultimately causing the land to succumb to a vacuum-like activity. Last year, Madera County saw an average drop of over 1 foot in land levels—with subsidence occurring at a rate of 18″ per year on the County's West side.[5] It is important to note that typical groundwater aquifers are recharged once a significant rain event occurs, but subsided land does not. It can be compared to a plastic bottle literally being vacuumed sucked dry—but unable to be refilled.

Madera County is the top part of the Friant Water System, managed largely by the Madera Irrigation District [MID] and second to that the Chowchilla Irrigation District. The Friant Division is the central piece of the Central Valley Project plan and irrigates more than 1 million acres on the valley's east side. Beginning at Millerton Lake and dammed by Friant Dam, water is diverted through the Friant-Kern Canals to southern counties including Fresno, Kings, and Kern. Diverting water west toward the dryer eastern Madera and Fresno areas is the Madera Cross Canal.[6]

The Central Valley Project [CVP], managed by the Bureau of Reclamation (Bureau), provided for the construction of Friant Dam in 1944. This Project set up the current system of exchanged water deliveries between the Sacramento and the San Joaquin Rivers. The Friant system's current practices of classification deliveries were also born from the CVP, specifically Class 1 and Class 2 water. Under normal conditions, 840,000 AF of northern California water is delivered to the Mendota Pool via the Delta-Mendota Canal for use by west side agencies with historic San Joaquin water rights[7]—known as the exchange contractors. As a result, 800,000 AF of water may be diverted for the Friant water users on the eastern valley floor—which is classified as Class 1 water. An additional 140,000 AF of water is available for Friant contractors if and when it becomes evident that the needs of the Class 1 water will be met by that year's water supply. This 140,000 AF is designated as Class 2 water. This year, the Bureau of Reclamation has determined that the supply

[1] Madera County Ag Commissioner's 2013 Annual Crop Report (online at *www.madera-county.com*).

[2] Central Valley Farmland Trust 2014 Central Valley Review *pp. 244–258.*

[3] Madera Irrigation District.

[4] Madera County General Plan Update 2013 *pp. 89–145.*

[5] Central California Irrigation District [CCID] and San Joaquin Exchange Contractors, *2013 Merced and Madera Subsidence Study.*

[6] Friant Water Users Authority, *Friant Division Facts 2014* (online *www.friantwater.org*).

[7] *See above.*

for Class 1 water is zero, and therefore, zero is also available for the Class 2 water users. This designation of zero is unprecedented and greatly impacts the future prosperity of not just Madera County agriculture, but the entire Central Valley. In addition, the Bureau's Operation and Maintenance costs have sky-rocketed to the local irrigation districts—as there is less water moving through the system bringing the cost per AF to astronomically high levels.[8] Even though the Friant users are receiving a zero allocation this year, they will be bearing a major portion of the O&M fees associated with the Central Valley Project in 2015.

Madera County Farm Bureau's membership is largely composed from Class 2 water users to the Friant system—to a much lesser extent Class 1. But it's a forgone conclusion at this point that most of our membership has been or is on the books to drill deeper wells in anticipation of this crisis. The waiting list for a well drill is over 13 months from the time of booking, and can exceed costs of $1 million. This figure—although staggering, is a far cheaper investment than losing highly productive almonds or pistachio orchards.

In addition to the raw economic affects this zero allocation of northern California/delta water brings to agricultural operators, the rural farming communities and labor services that go along with agriculture have been hit hard. Finishing the first quarter of 2014, due to lack of rain and available irrigation practices, nearly half of Madera County's temporary work force was left out of work or placed on temporary leave.[9] With no weeds to spray and any trees or vines to prune, Madera County faced a staggering increase in unemployment—from 11 percent on average to 26 percent.[10] Madera County's rural communities of Firebaugh and Mendota are slated to run out of municipal water by July this year.

Since the effects of the zero allocation to the Friant system by the Bureau of Reclamation have such far reaching consequences, the Madera County Farm Bureau is concerned that a full accounting of water supplies by the Bureau has not been made available. Some water continues to be made available to small rural towns that rely solely on Friant water for municipal purposes, understandably by way of a reserve called ''Health and Safety Water,'' that was produced by shortening the restoration flows dedicated in the San Joaquin River Restoration Program.[11] The MCFB was pleased that restoration activities were curbed in January 2014;[12] however **it is critical that the amount of water saved and the Bureau's dedication of its uses be published as soon as possible, least a request demanding such information from the Bureau and the Department of the Interior be necessary.**

The aforementioned model of Friant water user classification and its efficacy had never been tested in a manner that actually involved a zero water allocation from the Bureau. It had however—been heavily theorized in a model developed by the Technical Advisory Committee [TAC] to the San Joaquin River Restoration Project [SJRRP], developed by way of the San Joaquin River Settlement Agreement (Settlement Agreement/SA). The MCFB maintains a seat on the Board of Directors at the Resource Management Coalition [RMC], to which public and non-public presentations are made by the Bureau of Reclamation on the status of the SJRRP to a collective group of San Joaquin River stakeholders, the Exchange Contractors, State Department of Water Resources, the U.S. Fish and Wildlife Service, the National Marine Fisheries Service, and multiple irrigation districts. Throughout last year, the Bureau suggested through multiple reports[13] and letters from the State Water Resources Control Board [SWRCB] that a 0 percent allocation was impending based on hydrological models. The planning for this event is therefore derived to be a contingency of the SJRRP, and the MCFB is deeply affected by its implementation. To fully understand the nature of how the Settlement Agreement [SA] affects MCFB and its members, a summary of the settlements key provisions is necessary. The SJRRP is a direct result of a Settlement (known as the SA), reached in September 2006 on an 18-year lawsuit to provide sufficient fish habitat in the San Joaquin River below Friant Dam near Fresno, California, by the U.S. Departments of the Interior and Commerce, the Natural Resources Defense Council [NRDC], and the

---

[8] Madera Irrigation District 2014 *O&M Charges and Fees Schedule.*

[9] Madera County Economic Employment Department, *1st Quarter Economic Outlook, pp. 13–18.*

[10] *See above.*

[11] Madera Irrigation District *Friant System pp. 89,* San Joaquin River Restoration Program, *FEIR/EIS 2012 pp. 439.*

[12] Bureau of Reclamation, SJRRP, Press Announcement (online at *http://restoresjr.net/news/MP-14-012SJRRPCeaseFlows1MonthEarly.pdf*).

[13] *Draft Channel Capacity Report for 2014,* Bureau of Reclamation, presented at RMC *Permits 11885, 11886, and 11887 and License 1986 of U.S. Bureau of Reclamation,* Letter from SWRCB dated October 21, 2013.

Friant Water Users Authority [FWUA]. The settlement received Federal court approval in October 2006. Federal legislation was passed in March 2009 authorizing Federal agencies to implement the settlement.[14] The settlement is based on two goals:

Restoration: To restore and maintain fish populations in ''good condition'' in the main stem of the San Joaquin River below Friant Dam to the confluence of the Merced River, including naturally reproducing and self-sustaining populations of salmon and other fish.

Water Management: To reduce or avoid adverse water supply impacts to all of the Friant Division long-term contractors that may result from the Interim Flows and Restoration Flows provided for in the settlement.

The MCFB and its members are greatly and frequently affected by the SA's water management strategies—which are directly influenced by the SA's restoration objectives. These two goals are often contradictory in nature and in a case like this year's extreme drought, have made the SA un-implementable by the State, the Federal Government, and those locally involved.

By way of example, the SJRRP's efforts to build habitat required for the reintroduction of anadromous fish has stalled for multiple reasons—however the plan to support a small population of transplanted fish has moved forward—without any of the infrastructure required to keep the fish alive. This took a significant amount of water out of the system for the Class 1 and 2 Friant water users heading into a critical drought year. The information can be summarized by the Bureau's designated Restoration Administrator, Tom Johnson in the following manner:

''The winter of 2013–2014 is shaping up to be one of the driest in California history . . . the opportunity to conserve unreleased Restoration Flows to support the Restoration Program in the future and improve water supplies in the region in this incredibly dry year **was** a consideration . . . ultimately, it was the . . . consensus that an early reduction of flows, **while not biologically beneficial in its own right, is biologically reasonable** . . . given the anticipated sufficient water temperatures in critical areas of the river . . .''[15]

The MCFB contends that this practice, although discussed and determined legally under the confines of the SA, is a horrendous practice—effectively placing a non-existent population of fish over a very real and present population of people and agricultural businesses. The amount of water that was dedicated to the 2014 Restoration Flows was over 250,000 AF.[16] Although the MCFB appreciates that an overall ''ramp down'' of restoration flows occurred, this amount of water being dedicated to something that the Bureau's own panel of experts and scientists has claimed is pointless is a massive waste of water and precious wealth for the Central Valley.

The MCFB would like to offer a set of solutions to this water crisis, immediate and long term. These solutions have been tailored to the jurisdiction of this committee, the House Natural Resources Committee—and should be viewed through its ability to enact change through its jurisdiction.

IMMEDIATE WATER CRISIS SOLUTIONS

I. Expedition of Water Deliveries by Maximization of Through Delta Pumping

The need for expedited water deliveries—specifically throughout the delta and Mendota Pool is extreme and can be performed in real time. Achieving maximum flexibility in delta export operations will be key in allowing the Bureau to meet Exchange Contractor substitute water supply operations, which is critical for Friant to be able to use whatever supplies may be generated (or stored) in the upper San Joaquin River watershed.

Water deliveries are presently being hampered by an inadequate definition of what is considered a protected v. threatened species under the Endangered Species Act. This committee has the power to review and change this law to better define the nature of what an endangered species is AND the success criteria required for it to be delisted. This change, although controversial, may be considered to sunset by 2015, to at minimum allow some form of relief for farmers during this crisis.

This action would also bring the Tracy Pumping Plants back online at a greater capacity, providing much needed relief for the recirculation efforts on the San Joaquin River.

[14] Bureau of Reclamation, SJRRP (online *http://restoresjr.net/background.html*).

[15] San Joaquin Restoration Program, *Restoration Administrator Flow Recommendation,* January 31, 2014, *''Recommendations for 2014 Restoration Flows.''*

[16] *See above.*

In addition to these immediate fixes, any and all water dedicated toward cold water promotion in attempts to minimize turbidity throughout the Central Valley Project must cease immediately. This is a wasteful practice in the delta, given the drastic need for all the water available to supply people and people's food supply.

II. San Joaquin River Restoration Plan Amendments (Pub. L. 111–11)

The SJRRP provides for a dedicated "cold water fishery" on the San Joaquin River, based on historical hydrographic data and evidence of previous cold water activities nearly 100 years ago. It was this biome that the SJRRP seeks to reproduce in the present day environment in an attempt to bring back anadromous salmon numbers. However, there are numerous habitat necessities that will be required prior to implementing a cold water fishery—namely a high volume of water, side channel habitat construction and spawning gravel implementation, which at this time make this condition in the SJRRP unworkable.[17] This committee has the jurisdiction to revisit Pub. L. 111–11, and develop a more logical timeframe for which to implement these restoration objectives—but moreover, to delay any activities associated with it implementation in the next year—based on the critical water year. The SJRRP's goal of implementing restoration should also be based on minimizing a waste of taxpayer dollars as well as facilitating water deliveries to the Friant system.

Again, this action can be considered to sunset by 2015, to at minimum allow some form of relief for farmers during this crisis.

LONG TERM WATER CRISIS SOLUTION

I. Investment in Water Storage Infrastructure

One of the greatest and most imperative solutions for long term drought crisis aversion is the development of storage throughout California. For MCFB members—and for most within the Friant system, the development of a storage facility in the upper San Joaquin River Basin (Project) would provide massive amounts of direct relief for 5 counties (Madera, Merced, Fresno, Kings, and Kern), more than 6 million acres of irrigable Ag land, and over 1 million people. This is a bold statement, but upon elaboration more can be derived from its roots.

—Upper San Joaquin Storage Site has already been authorized by Congress[18]
—Project does not touch the delta or is hindered by through-delta conveyance
—Project is the strongest contender for a local cost share—not also requiring/needing a State cost share component
—Local irrigation districts will not or don't have to be required to pay for project

This storage site, colloquially known as Temperance Flat, regardless of the end use or ownership—is the only one in the cue that has the ability to bring water into the San Joaquin River system *directly*. This means that should the end purpose of the near 500,000 AF generated by the Project.

The Bureau of Reclamation, in its January 2014 Feasibility Report,[19] cited that the potential net effects of a storage project in the upper San Joaquin would, "significantly contribute to the success of flow and therefore the success of a Chinook salmon population, known to be affected by water temperatures . . ." The MCFB views this benefit—although not directly benefiting farmers, as an overall benefit of the project thus contributing to more water system wide.

In summary, the drought crisis has been influencing catastrophic effects on members of the MCFB. We are estimating a total net loss of $65 million in crop damage, $455 million in our labor forces, and nearly $275 million lost due to water lost on the exchange market. We hope that this committee, through its jurisdiction can enact the immediate and long term solutions we've proposed.

The Madera County Farm Bureau appreciates the opportunity to provide testimony today. We have included a letter from our neighboring Farm Bureau, Tulare, to be included as part of the record.

The CHAIRMAN. Thank you very much, Mr. Coleman.

[17] SJ Settlement Agreement, Case 2:88–cv–01658–LKK–GGH Document 1341–1 Filed 09/13/2006 Page 13 of 80.

[18] *Pub. L. 108–7, Division D, Title II, Section 215, Omnibus Appropriations Act 2/2003.*

[19] Draft, *Upper San Joaquin River Basin Storage Investigation*, Feasibility Report, January 2014.

Our next invited witness is the Honorable Felicia Marcus, who is the Chairwoman of the State Water Resources Control Board of California.

Chair Marcus is not here, and I have to say, as the Chairman of the committee, I am really disappointed because we invited her probably a week-and-a-half ago to testify. Obviously, this is a very important issue, and it seems to me from my perspective that the Water Control Board, at least the Chair, ought to be here to testify. We are here to get information.

So we invited her probably a week-and-a-half ago, and it was only yesterday that we got a letter saying that she was not going to show up. Now, I find it very ironic. On the front page of the paper this morning, the State Water Control Board was 45 minutes away in Firebaugh, and yet the Chair of the Water Control Board couldn't show up today. So I am very, very disappointed that she is not here, and I just simply wanted to state that for the record.

So next I want to recognize Mr. David Murillo, who is the Regional Director of the Mid-Pacific Region of the U.S. Bureau of Reclamation.

Mr. Murillo, thank you for being here. You are recognized for 5 minutes.

STATEMENT OF DAVID MURILLO, REGIONAL DIRECTOR, MID-PACIFIC REGION, BUREAU OF RECLAMATION, U.S. DEPARTMENT OF THE INTERIOR

Mr. MURILLO. Thank you, Mr. Chairman. Chairman Hastings and members of the committee, I am David Murillo, Regional Director of the Mid-Pacific Region for the Bureau of Reclamation. I am pleased to be here today alongside our partners to describe the actions that are underway to address the drought in California. My full written statement has been submitted for the record.

We are all aware of the severity of this drought, so I will skip straight to the discussion of what Reclamation is doing to help both at the operational level and within our budget.

First, Reclamation and our stakeholders are implementing a demonstration project for managing Old and Middle River, or OMR, flows in the delta. Basically, the demonstration project will improve operational stability and result in more efficient CVP and State project operations. This demonstration project is being implemented in 2014 and will be reevaluated for 2015 operations.

Second, in January, Reclamation worked with DWR on a Temporary Urgency Change Petition that was submitted to the State Board on January 31. The petition requested a reduction in the delta outflows required by State Water Rights Decision 1641, as well as other actions to maintain delta salinity requirements. The State Board issued an order in response to the petition, and we requested to extend the January order through the end of March.

Third, as described in our 2014 CVP Water Plan, Reclamation proactively requested an early determination from NMFS of the San Joaquin River inflow-to-export ratio based on the January runoff forecast and the predicted continuing dry February forecast results based on the need to plan in advance and provide some certainty in operations to accommodate water transfers.

And fourth, we have been working collectively with our contractors to develop environmental documents to support water transfers should conditions allow and sellers are willing to make water available. Just last week, we publicly released two transfer alternative documents. First transfer water from north-of-delta contractors to south-of-delta contractors; and two, transfers of non-project-based supplies from Sacramento River Settlement contractors to in-basin buyers north of the delta. Cumulatively, these alternatives could make 100,000 to 200,000 acre feet of water available.

As we move forward in this drought year, Reclamation, DWR, NMFS, the Fish and Wildlife Service, Department of Fish and Wildlife, Federal and State contractors, and the State Board are working to develop an operational plan for the remainder of the water year which will serve as a contingency plan under the NMFS BO. This plan outlines assumptions for all users in California water to plan for and implement drought response measures, as necessary.

Turning from operations to the funding perspective, we have worked for years to maximize the budgetary resources available for water supplies in California. Every year for the past two decades, hundreds of millions of dollars in Federal resources have been provided annually in this State, much of it here in the Central Valley, to develop new water supplies, maximize conservation, improve existing infrastructure, and finalize innovative agreements among water users. As the other witnesses here can attest, local communities in California are on the forefront of water supply efficiency and modernization of their delivery systems. These are summarized in my written statement.

Of course, there is always more to do. We remain committed to longer term solutions that will create a more sustainable future for the CVP. We are pressing forward on the feasibility studies for new and expanded reservoir storage in the Central Valley. We have completed four major reports on storage projects since July of last year. Specifically, we released a Draft Environmental Impact Statement for the Shasta investigation, and in December we released a draft appraisal report on the expansion of San Luis Reservoir, as well as a progress report for the north-of-delta off-stream storage. Then in February we released the Draft Feasibility Report for the Upper San Joaquin River Basin Storage Investigation. And we are planning to release the Draft Environmental Impact Statement for it this year. Last, we expect to complete the Final Feasibility Report and Final EIS for Shasta by the end of this year as well.

In closing, I thank the committee for its attention to this issue, and for fair consideration of all that we are doing to operate the State and Federal projects in compliance with the law. Reclamation values its working relationship with all the parties represented here today. I would be glad to answer any questions at the appropriate time. Thank you.

[Applause.]

[The prepared statement of Mr. Murillo follows:]

PREPARED STATEMENT OF DAVID MURILLO, REGIONAL DIRECTOR, MID-PACIFIC REGION, BUREAU OF RECLAMATION, U.S. DEPARTMENT OF THE INTERIOR

Chairman Hastings and members of the committee, I am David Murillo, Regional Director of the Mid-Pacific Region for the Bureau of Reclamation (Reclamation). I am pleased to represent the Department of the Interior (Department) today, alongside our partners including the State of California and the water community, to describe the actions that are underway to address the drought in California.

As the committee is acutely aware, California is experiencing its most severe drought in recent history. We are now more than two-thirds of the way through the rainy season and many areas of the State are 60 to 75 percent below average annual precipitation totals for this date. It would take more than ½ inch of rain from Redding to Fresno every other day until May to get back to average precipitation, and even with such precipitation, California would remain in drought conditions due to low water supplies in reservoirs from the two previous dry years. Despite recent storms, our very low reservoir and snowpack levels dictate that we must plan ahead and conserve more water. Reclamation, the State, and our Federal partners have not been standing still waiting for this drought to develop. State and Federal water managers are working hand in glove in a delicate balancing act to optimize water allocations, both short-term and long-term. For my testimony, I would like to summarize some of our actions at the operational-level aimed at reducing the impacts and optimizing the use of existing water supplies this year, and then I will move on to some of the funding issues relevant to this discussion.

First, Reclamation and multiple stakeholders developed and are now implementing a demonstration project for managing Old and Middle River [OMR] flows in the delta. The demonstration project will use a ''flow index'' that can be calculated in real-time to make decisions instead of tidally filtered gauge data that can take days for determining OMR flow requirements associated with the U.S. Fish and Wildlife Service [USFWS] and National Marine Fisheries Service [NMFS] Biological Opinions [BOs]. Implementing the OMR Index Demonstration Project will improve operational stability and simplify accounting for the many factors affecting OMR flow, and result in simplified and more predictable Central Valley Project [CVP] and State Water Project [SWP] operations. This demonstration project is being implemented in 2014 and will be reevaluated for 2015 operations.

Second, in January, Reclamation worked with the California Department of Water Resources [DWR] to develop a Temporary Urgency Change Petition that was submitted to the State Water Resources Control Board (State Board) on January 31, 2014. The Temporary Urgency Change Petition requested a reduction in the delta outflows required by State water rights Decision 1641, as well as other actions to maintain delta salinity requirements. The State Board issued an Order in response to the petition on January 31, 2014. In late February, Reclamation and DWR requested the State Board to extend the January Order through the end of March. The State Board granted this extension on February 28, 2014. As part of the Temporary Urgency Change Petition and Order, Reclamation is providing support to USFWS and the California Department of Fish and Wildlife [DFW] to perform additional delta smelt and salmon monitoring. This monitoring is providing additional information and will provide information more quickly on fish movement and presence to inform operations under the Temporary Urgency Change Petition and Order. Reclamation and DWR will continue to monitor hydrologic conditions to determine whether additional drought response actions should be requested from the State Water Board. Also, Reclamation is actively monitoring the State Board's flow and salinity standards on the San Joaquin River at Vernalis. If necessary, Reclamation and DWR may be requesting a relaxation of Vernalis salinity standards to conserve water in storage that can be used later to ensure our ability to keep control over delta salinity over the long term, should the drought continue.

Third, as was detailed in our 2014 CVP Water Plan, Reclamation has taken a number of steps to facilitate water transfers. Reclamation requested an early determination from NMFS of the San Joaquin River inflow to export (I:E) ratio requirement based on the January runoff forecast and the predicted continuing dry February forecast results. To allow water users to plan in advance and to provide some certainty in operations to accommodate water transfers, on February 7, NMFS agreed to establish an I:E ratio of 1:1 for April–May, 2014 earlier in the year than they normally would so that we could plan for less restrictive CVP and SWP exports. Also consistent with the 2014 Water Plan, Reclamation has been working collaboratively with its contractors to develop environmental documents to support water transfers, should conditions allow. During the Week of March 10, we publically released two transfer alternatives: (1) transfer of water from north of the delta contractors to south of the delta contractors; and (2) transfer of non-project base supplies from Sacramento River Settlement Contractors to in-basin buyers north of the delta. Cumulatively, these alternatives could make 100,000 to 200,000 acre-feet of water available to those most in need. In addition, Reclamation has been working closely with the DWR and the State Board to facilitate water transfers.

As we move forward in this drought year, Reclamation, DWR, NMFS, USFWS, DFW, Federal and State contractors, and the State Board are working to develop an operations plan for the remainder of the water year, which will serve as a contingency plan under the drought exception procedures in the NMFS BO. This plan will outline operations and assumptions (allocation, refuges, barriers, cold water pool, water quality, fisheries, and the possibility of entering into another drought year in 2015) to allow all agencies and users of California water to plan for and implement drought responses measures as necessary.

Through these and other actions, Reclamation is working closely, on a day-to-day basis, to coordinate and communicate proactively with the State of California and within the Federal family. High level leadership calls are being held weekly to identify issues before they become problems, and to find solutions to provide water for our customers and protect irreplaceable natural resources.

In November 2013, the administration launched the National Drought Resilience Partnership [NDRP] to help communities better prepare for droughts and to reduce impacts on families and businesses. The NDRP is coordinating Federal efforts across the country and working closely with State and local governments and other partners to improve community preparedness and resilience to drought. With the severe drought in California, the NDRP is also playing a critical role in response, helping to connect communities to the Federal assistance they need.

Turning from operations to funding, we have worked for years to maximize the resources available for water supplies in California. Every year for the past two decades, hundreds of millions of dollars in Federal resources have been provided annually in this State, much of it here in the Central Valley, to develop new water supplies, maximize conservation, improve existing infrastructure and finalize innovative agreements among water users. As the other witnesses here can attest, many local communities in California are on the forefront of water supply efficiency and modernization of their delivery systems. The President's visit to the Central Valley last month, as well as the Secretary Jewell's visit last week, made clear that the administration understands the seriousness of the situation here. Two weeks before the President's visit, our previous Commissioner, Mike Connor, came to Sacramento to announce a 2014 funding opportunity of up to $14 million in Federal assistance for irrigation districts, water districts, tribes and other water or power entities to cost share on projects that create new supplies for irrigation and improve water management. This opportunity is part of a partnership between Reclamation and the Natural Resources Conservation Service [NRCS], whereby NRCS will provide funding and technical assistance for on-farm projects such as tail water recovery systems, conversion to sprinkler or drip systems, and micro-irrigation investments. Reclamation and NRCS will each provide up to $7 million for this effort. The deadline for submitting proposals is Monday, March 24 at noon, and we anticipate project selections will be announced by late May or early June.

These efforts are not new in the Mid-Pacific Region. Since 2009, Reclamation has provided over $42 million in financial assistance to water purveyors in the Region for agricultural and urban water use efficiency improvement/management projects. Through various programs such as CALFED, Bay-Delta Restoration Program (NRCS Partnership), WaterSMART, and the Water Conservation Field Services Program, combined with recipient cost share, over $138 million has been invested in water efficiency improvement projects over the last 4 years. Collectively these projects conserve approximately 274,000 acre-feet of water annually and have been proven as one of the most cost effective ways to increase the available supply of water in California, and elsewhere. Through the title XVI water reuse program alone, municipalities throughout California are now making use of approximately 350,000 acre-feet of recycled water annually, reducing reliance on the over-allocated Bay-Delta and Colorado River systems.

Reclamation recognizes the need to fund projects that address water supply sustainability and stretch limited water supplies. This is made all the more relevant when you consider that the hundreds of thousands of acre-feet of CVP water that was rescheduled from 2013 into 2014 is proving crucial to providing water supplies this year. For some districts, this water is their only source for 2014 supplies. A significant amount of this rescheduled water would not have been available without the conservation investments made with our partners in years past under these programs.

In addition, the projects that were funded in 2009–2011 by the American Recovery and Reinvestment Act under the authority of the Reclamation States Drought Relief Act of 1991 have now been implemented. Reclamation provided $40 million in funding in the Sacramento and San Joaquin Valley for well rehabilitation, new wells, and temporary pumps and pipes. This new infrastructure is providing a water supply to areas that previously did not have access to a supply and is assisting growers to be more resilient to drought.

It has been 2 years since Reclamation and the San Luis and Delta-Mendota Canal Authority completed construction of a 500-foot connection between the State and Federal projects just west of Tracy. The Delta-Mendota Canal/California Aqueduct Intertie addresses conveyance conditions that had restricted use of the Jones Pumping Plant to less than its design capacity, potentially restoring 35,000 acre-feet of average annual deliveries to the CVP. The Intertie provides redundancy to portions of the State and Federal water distribution system, allows for maintenance and repair activities that are less disruptive to water deliveries, and provides the flexibility to respond to CVP and SWP emergencies. In the first 18 months of operation, nearly 73,000 acre-feet of additional CVP water was pumped through the Intertie. It was a successful project, and is illustrative of the working relationship we have with the State and our water contractor community.

Of course, there is always more to do. We know there will be more tough choices to maintain basic supplies if a fourth straight dry year materializes. Various Federal and State agencies are assessing the amount of "carry-over" supplies that must be retained in our reservoirs to maintain salinity control in the delta to ensure that it can continue to be used as a water supply source and to provide for health and safety purposes in case of a fourth straight dry season, and this possibility will inform our thinking for the rest of 2014.

Although we are focused on near-term actions to address the drought, we also remain committed to finding longer term solutions that will create a more sustainable future for the CVP. We continue to press forward on the feasibility studies that examine the potential for new and expanded reservoir storage in the Central Valley. Of note, we have completed four major reports on storage projects since July last year. Specifically, in July we released the Draft Environmental Impact Statement for the Shasta Lake Water Resources Investigation, and in December we released a Draft Appraisal Report on the expansion of San Luis Reservoir, as well as a progress report for the North-of-Delta Offstream Storage Investigation. Then in February we released the Draft Feasibility Report for the Upper San Joaquin River Basin Storage Investigation. In addition, Reclamation is planning to release the Draft Environmental Impact Statement for the Upper San Joaquin Investigation and complete the Final Feasibility Report and Final Environmental Impact Statement for the Shasta Lake Water Resources Investigation by the end of this year. Reclamation, through the San Joaquin River Restoration Program, is also supporting the development of groundwater recharge projects in support of the water management goals of the Program.

Finally, I would note that for the long-term, the administration remains committed to working closely with the State of California to achieve the co-equal goals of (1) improving California's water supply reliability; and (2) protecting, conserving, and restoring the bay-delta environment. In addition to the water management measures discussed above, we continue to work in close partnership with the State in developing the Bay-Delta Conservation Plan.

In closing, I thank the committee for its attention to this issue, and for fair consideration of all we are doing to operate the State and Federal projects in compliance with the law for the benefit of all Californians and the environment. Reclamation values its working relationship with all the parties represented here today. I would be glad to answer questions at the appropriate time.

The CHAIRMAN. Thank you very much for your testimony.

Our next witness is Ms. Janelle Beland, who is the Undersecretary of the California Natural Resources Agency in Sacramento.

For the record, our invitation was to Mr. John Laird, who is the Secretary. He couldn't make it, and he was gracious enough to send his Undersecretary, and we thank you very much for being here.

I just wish the same courtesy could have been given to us from the State Water Resources Control Board.

Ms. Beland, you are recognized for 5 minutes.

STATEMENT OF JANELLE BELAND, UNDERSECRETARY, CALIFORNIA NATURAL RESOURCES AGENCY, STATE OF CALIFORNIA

Ms. BELAND. Good morning, Mr. Chairman and members of the committee. I am Janelle Beland, Undersecretary for Natural Resources for the State of California. The State appreciates the invitation to appear before this committee today and offer testimony on our response to the current drought.

With California now in our 3rd dry year of weather and shrinking reservoir supplies, we are reminded once again that nothing focuses California's attention on our limited water resources like drought. California is experiencing a severe drought of uncertain duration. On the heels of two previous dry years, storage in the State's major reservoirs and the water content of the Sierra snowpack remain well below average for the date.

Recent storms have not ended the drought, and the window for California to gain significant precipitation is closing. The latest National Weather Service data continue to show nearly the entire State in severe drought and nearly two-thirds in extreme drought.

Federal, State, and local water projects that rely on snowpack in the Cascades and the Sierra, the source of nearly one-third of California's developed water supply, will be operating under unprecedented dry conditions this summer, and we will be challenged to manage our system to conserve vital reservoir storage.

Typically, the Sierra Nevada snowpack melts in the spring and summer. It collects in reservoirs to provide about one-third of the water Californians use each year. Some communities are running low on drinking water. Many of California's rivers and streams are also running very low, and farmers who rely on surface water for irrigation are faced with difficult decisions to plant crops amidst great uncertainty about whether State and Federal water infrastructure will be able to deliver the water needed to supplement local supplies to grow their crops.

Everyone who relies in whole or in part on project water—farmers, fish, people in cities and towns—will get less water this year. Simply put, there is not enough water to go around, so we need to conserve and make some strategic decisions now, planning for the worst if we do not get much more precipitation in the next few weeks.

One of the most important lessons learned from our previous record dry years, such as 1976 and 1977, is that delay only makes the effects of the drought worse. Just like the Governor has asked all Californians to conserve water around their homes, we are taking the same actions for the State on a much larger scale.

On January 17, Governor Brown issued an emergency Proclamation of Drought. That proclamation ordered that our Department of Water Resources work constructively with fellow State and Federal agencies to take proactive steps now to preserve our ability to manage water supplies across a broad array of needs should this drought worsen. The Governor's proclamation is the fourth action taken by a Governor since 1987 to deal with drought on a statewide basis.

The California Department of Water Resources runs the State Water Project. With Lake Oroville and the California Aqueduct that winds across two-thirds of the State, the State Water Project delivers water pumped from the delta to 25 million Californians and 750,000 acres of farmland.

One primary concern from the State's perspective has been to ensure that enough water can be directed to communities for basic needs such as drinking water and water for sanitation and firefighting. While some communities have adequate water supplies saved locally for such purposes, other communities need continued exports from the delta for these essential purposes.

Another primary concern is the need to prevent salt water intrusion into the interior delta, where a large portion of the State's fresh water supplies are conveyed for human and agricultural use. A certain amount of outflow must continue throughout dry months to push back salt water from the interior delta. If there is not enough water to maintain this balance throughout the year, fresh water sources traveling through the interior delta will become contaminated. This is a very real concern this year.

In recognition of both of these concerns, on January 29 the Department of Water Resources and the Bureau of Reclamation asked the State Water Board to adjust water rights permit and license terms that normally control State Water Project and Federal Central Valley Project operations. DWR and the Bureau sought this temporary urgency change in order to preserve dwindling supplies in upstream reservoirs for farms, fisheries, and cities and towns as this drought continues.

The relief sought in the petition would also provide additional time to assess how much water the projects would need for salinity control and basic needs such as drinking water and sanitation throughout this year.

This temporary urgency change did the following: it allowed a reduced delta outflow so that the State and Federal projects could conserve their dwindling supplies in reservoirs for later in the year; it allowed for the operation of the Delta Cross Channel gates in real time so less flow would be needed to repel salinity; and it established a real-time drought operations management team. Without these changes, various regulations would require us to release water in our reservoirs now. The amount usually required to be released from reservoirs this time of the year was set assuming a dry year, but not a drought of this magnitude.

The CHAIRMAN. Could I ask you to summarize your final part?

Ms. BELAND. Sure.

The CHAIRMAN. Your full statement will appear in the record.

Ms. BELAND. Yes, I get it.

I guess on that, I will just wait for questions. Thank you very much.

[The prepared statement of Ms. Beland follows:]

PREPARED STATEMENT OF JANELLE BELAND, UNDERSECRETARY, CALIFORNIA NATURAL RESOURCES AGENCY, STATE OF CALIFORNIA

OVERVIEW

Good morning Mr. Chairman and members of the committee. I am Janelle Beland, Undersecretary of the California Natural Resources Agency. The State of California appreciates the invitation to appear before your field hearing today to offer testimony on our response to the ongoing drought crisis.

California is experiencing a severe drought. On the heels of 2 previous dry years, all of the State's major reservoirs remain well below average storage for the date. Statewide, the water content of the Sierra snowpack also is well below average for the date. Recent storms have not ended the drought, and the window for California to gain significant precipitation is closing. The latest National Weather Service data continue to show nearly the entire State in severe drought and nearly two-thirds in extreme drought.

Although long-range forecasts suggest a shift to weak El Niño conditions in the coming months, this does not mean that the drought in California will be over next winter. State and Federal water project operators and environmental and water quality regulators are working together in real time to exercise as much flexibility as possible under regulatory standards to allow for the capture and storage of water from the Sacramento-San Joaquin Delta (Delta). Every effort has been made to maximize the amount of water the projects could export during the storms in February and March, with the realization that this may be the last opportunity to capture and store unregulated flow during this winter season.

These efforts are being closely coordinated with State and Federal fishery agencies and the State Water Resources Control Board (State Board), which is exercising flexibility allowed under the law. This real-time water management will continue to adjust operations in response to changing conditions.

In the coming weeks, in order to help preserve water supplies in upstream reservoirs and limit salinity encroachment in channels of the Delta, the California Department of Water Resources [DWR] is developing plans to install temporary emergency rock barriers across three Delta channels. DWR is working with Federal and State wildlife agencies and the U.S. Army Corps of Engineers to gain permits for installation of these emergency barriers, which would be removed in the fall. Similar emergency barriers were last installed in the drought year of 1977, and the barriers worked as intended to help control salinity.

OVERVIEW OF STATEWIDE CONDITIONS

As previously mentioned, despite several consecutive days of rain across California this month, we are significantly behind average precipitation conditions for this time of year.

The statewide snowpack shows 29 percent of average snow water content for the date, slightly less than last week's measurement. The snowpack water content is currently at 19 percent in the northern Sierra, 35 percent in the central Sierra, and 33 percent in the southern Sierra.

Typically, the Sierra Nevada snowpack melts in the spring and summer. It collects in reservoirs to provide about one-third of the water Californians use each year.

Major reservoir storage rose slightly over the last couple of weeks but is still significantly below average. Shasta stands at 58 percent of typical storage for this time of year. Oroville storage is 62 percent of average and Folsom Lake is at 67 percent of average storage for this time of year.

Federal, State, and local water projects that rely on snowpack in the Cascades and the Sierra—the source of most of California's developed water supply—will be operating under unprecedented dry conditions this summer, and will be challenged to manage their systems to conserve vital reservoir storage.

The very low reservoir and snowpack levels dictate that we must be prudent with our minimal water supplies, and that requires additional flexibility to operate the State and Federal water projects. In this extraordinarily dry year, all water users, including agricultural, municipal, and fish and wildlife uses, will be impacted.

To maximize flexibility, the project operators, DWR and the U.S. Bureau of Reclamation (Bureau) have coordinated closely to exercise maximum flexibility and allow the water projects to conserve and store water as they continue to assess the water needs for later in the year and into 2015. The California Department of Fish and Wildlife [DFW] and the National Marine Fisheries Service [NMFS] have coordinated closely with these agencies, and have worked to ensure that water management decisions do minimal harm to endangered and protected species.

One primary concern has been to ensure that enough water can be directed to communities for human health and safety purposes, which includes basic needs such as drinking water and water for sanitation and firefighting. While some communities have adequate water supplies saved locally for such purposes, other communities need continued exports from the Delta for these essential purposes. It should be noted that the agencies' intent has been to ensure enough water in communities for these essential purposes, but not to deliver exports for all normal usage (such as exterior landscape irrigation).

Another primary concern is being able to prevent saltwater intrusion into the interior Delta where a large portion of the State's freshwater supplies are conveyed for human and agricultural use. A certain amount of flow must continue throughout dry months to push back saltwater from the interior Delta. If there is not enough water to maintain this flow throughout the year, we will lose control over salinity in the Delta and fresh water sources traveling through the interior Delta will become contaminated. This severely compromises the water projects' ability to deliver water for basic public health and safety or irrigation uses. This is a very real concern this year.

January is typically the wettest month in California, but January 2014 proved an extraordinary anomaly. Scant rain or snow fell across the State for the entire month. It was the driest month on record for most places in the State, and it followed two previous dry years.

On January 29, 2014, DWR and the Bureau asked the State Board to adjust water rights permit and license terms that normally control State Water Project and Federal Central Valley Project operations. DWR and the Bureau sought this "temporary urgency change" in order to preserve dwindling supplies in upstream reservoirs for farms, fisheries, and cities and towns as the drought continues. The relief sought in the petition would also provide additional time to assess how much water the projects would need for salinity control and essential public health and safety needs throughout the year as mentioned above.

This temporary urgency change order (order) did the following:

- Allowed a reduced Delta outflow so that the State and Federal water projects could conserve their dwindling supplies in reservoirs for later in the year;
- Allowed for the operation of the Delta Cross Channel gates in real time so less flow would be needed to repel salinity;
- Established a Real Time Drought Operations Management Team.

DWR and the Bureau are now in the process of quantifying those public health and safety needs and defining precisely how any water set aside for public health and safety purposes may be used. The definition will not include deliveries to farms for irrigation or homeowners for lawn-watering.

On January 31, the State Board's Executive Director also advised that "junior priority" water-right holders may be ordered to curtail their diversions from the Sacramento and San Joaquin river systems. These curtailments would be structured to occur in a manner that respects water rights, with senior water right holders being the last to have their water restricted, as required under State law. Curtailments will be based on river gauges on each watercourse, and have not yet occurred.

Also, on January 31, DWR announced that its customers—29 public water agencies serving cities and farms—should expect no deliveries in 2014 if significant precipitation did not occur in the next few months. These customers could expect delivery only of "carryover" supplies that they had not used in 2013. The zero allocation is the first-ever for all customers in the State Water Project's 54-year history.

The announcement does not mean that anyone's tap will run dry, but it will trigger difficult decisions for many farmers, and it underscores the need for aggressive conservation by all Californians.

Also on January 31, DWR notified long-time water rights holders in the Sacramento Valley that their deliveries from the State Water Project may be cut 50 percent, the maximum cut permitted under contract, depending upon the results of future snow surveys. All of these settlement contractors are agricultural irrigation districts.

On February 7, the order was amended to include the flexibility needed after recent storms, when natural flows are high enough, so that the limits on Delta exports would not be in effect and normal conditions would apply.

Two separate, moderate storm systems moved across California in February. Water project operators worked with the State and Federal wildlife agencies to maximize regulatory flexibility so that as much storm runoff as possible could be captured and stored in San Luis Reservoir, south of the Sacramento-San Joaquin Delta, with minimal harm to Delta water quality and threatened and endangered Delta species. February 2014 proved wetter than January, but not enough to end the drought or avoid a high-stakes balancing act in water project operations.

Around February 9, Delta outflow started spiking after the first significant rain event to hit this winter. From February 10 to February 11, the State and Federal water projects increased their pumping from the Delta to about 6,000 cfs, maintaining that level until February 18.

During the month of February, the State and Federal projects received additional flexibility in the amount they could export from the Delta, via the coordination established under the Real-Time Drought Operations Management Team process for operating Delta facilities established by the January order. Additionally, Federal and State fish and wildlife agencies have made similar adjustments to export limitations based on their authorities and permits.

As a result, additional water was pumped from the Delta in February and March due to regulatory flexibility granted the projects by Federal and State fishery agencies. The rest of the water was pumped from the Delta under compliance with existing regulations that did not require use of the "temporary urgency change" or the easing of any standards designed to protect water quality and fish species.

Here is a more detailed analysis of Delta water project operations from February to today:

From Feb. 1 through Feb. 9, record-dry conditions in northern California kept Delta outflow (the volume of water flowing out of the Delta into San Francisco Bay) at roughly 7,000 cubic feet per second [cfs]. The combined export of the State Water Project and Central Valley Project (the amount of water diverted from the Delta into storage at San Luis Reservoir) was held at slightly under 1,000 cfs due to degraded water quality conditions in the Delta.

Around Feb. 9, Delta outflow started spiking after the first significant rain event to hit this winter. In addition, the Delta Cross Channel Gates were opened as part of the "temporary urgency change" order granted by the State Board. That order allowed for modified implementation of the requirements in D–1641, a water rights decision of the State Board that sets salinity and other water quality objectives for the Delta and Bay.

With Delta water quality improving, from Feb. 10 to Feb. 11, the combined water project exports ramped up to about 6,000 cfs and stayed there until Feb. 18.

By Feb. 18, Delta outflow had dropped to 8,000 cfs as the storm runoff dwindled.

Exports ramped down and reached the minimum health and safety level of 1,500 cfs by Feb. 23. Exports stayed at that minimum health and safety level until March 2. By March 2, with the return of significant rain, Delta outflow jumped to 26,000 cfs. Exports increased gradually from March 2 through March 4 to reach 6,000 cfs, then climbed over the next several days to reach 6,800 cfs.

Beginning on March 11, Delta outflow dropped to 11,000 cfs, then rose again to just under 17,000 cfs. Delta outflows are now headed back down to levels below 10,000 cfs by the middle of next week.

Combined water project exports have remained at just under 7,000 cfs and were scheduled to remain at that level through last weekend. The upper levels of exports were constrained by a multitude of existing regulations established to protect Delta fisheries including requirements of D–1641 and Federal rules to protect Delta smelt and chinook salmon, which are listed under the U.S. Endangered Species Act [ESA] and the California Endangered Species Act [CESA].

Operators of the State Water Project and Central Valley Project, together with representatives of the Federal and State fishery agencies and the State Board, are working collaboratively to find flexibility in the implementation of these regulations. They seek to maximize exports to the extent possible under the law, with the realization that the last set of storms may be the last opportunity to capture and store unregulated flow during this winter season.

In a normal year, the State and Federal water projects would be required to keep Delta outflow at 11,000 cfs during March, primarily to protect habitat for fish and wildlife. In particular, these flow requirements are elevated during this time of year because of the migratory life cycle of salmon. Increased flows help them move through the Delta ecosystem. This means that under a normal year's rules, 11,000 cfs is required to flow out of the Delta and through the San Francisco Bay. However, amid drought conditions and the need to conserve and export precious water, flexibility for this requirement was explored and the State and Federal fisheries agencies have advised that reducing the flow below 11,000 cfs will not unreasonably affect fish and wildlife.

The upper levels of exports are constrained by regulations to protect Delta fisheries and Federal rules to protect delta smelt and chinook salmon, which are listed under the ESA and CESA. Operators of the Federal and State projects are working collaboratively with the Federal and State fish agencies and the State Board to maximize flexibility in the implementation of these regulations.

During the coming weeks and months, the project operators will work in close coordination with the State and Federal fish agencies to ensure that the system stays within current requirements for fish. And they will closely monitor fish species affected by project operations to assess whether further protections are warranted.

DWR and the Bureau are gathering data and looking at how much water will be needed through the dry months and possibly into next year to maintain salinity control in the Delta, meet minimal public health and safety needs, and abide legal requirements to protect threatened and endangered fish. This review should clarify the water needed for these purposes for the rest of the year and possibly into next year if we experience a 4th dry year.

At that time, the State and Federal water projects will be able to update their allocation projections to their water contractors, based on the prudent assessment of water in the system for carryover storage needs for the coming dry months.

RECENT STATE ACTIONS

State and Federal water management agencies continue to work together to allow exports of additional water from the Delta based on storms in the last 6 weeks. Recent precipitation has provided a window of opportunity to capture additional water for storage both north and south of the Delta, and both the State and Federal water projects have increased pumping for a limited time to capture as much water as possible under current regulatory standards.

Last week, DFW and the State Board announced that they will expedite approval of storage tanks built by rural residents for domestic water use. These storage tanks help protect drinking water supplies and increase fire safety by giving rural residents a water supply that they can manage on their own property.

This week the State Board approved reduced cost financing for recycled water projects to speed the construction of such projects. They are also working to expedite approval of such projects.

DFW, USFWS, and NMFS last week released a contingency plan for the release of small fish raised in Federal and State hatcheries. Due to the drought, new measures will be taken to release the hatchlings in portions of the Delta that allow for their migration to the ocean while enabling their eventual return to lay eggs and continue their life cycle.

The Governor's Office and State agencies have launched *drought.ca.gov,* which will provide a central location for drought information. Agencies will continue their own drought Web pages, and *drought.ca.gov* will include a listing of these Web pages.

The Governor's Office of Planning and Research has posted online its drought toolkit for local governments, which outline actions that communities can take to respond to the drought.

This week on March 21, the Governor's tribal advisor will hold a statewide consultation call with tribal leaders to continue discussions on drought response with Interagency Drought Task Force officials.

The Department of General Services held a water conservation training last week for facility managers from State and local governments, as well as school districts across the State, to provide information and support to their water use reduction efforts. Over 300 managers from across the State participated.

The Governor's Office of Emergency Services continues to gather drought-related costs from State agencies and local governments, which is reported weekly to the Drought Task Force. The task force continues to meet daily to take actions that conserve water and coordinate State response to the drought.

On March 3, Governor Brown signed a $687.4 million drought relief plan (SB 103 and SB 104). Highlights of the package include:

- Accelerated grant expenditures of $549 million under Proposition 1E and Proposition 84 in the form of infrastructure grants for local and regional projects that are already planned or partially completed to increase local reliability, including recapturing of storm water, expanding the use and distribution of recycled water, enhancing the management and recharging of groundwater storage and strengthening water conservation.
- Thirty million dollars from the Greenhouse Gas Reduction Fund to DWR for direct expenditures and grants to state and local agencies to improve water use efficiency, save energy, and reduce greenhouse gas emissions from state and local water transportation and management systems.
- Fourteen million dollars for groundwater management across the State, including assistance to disadvantaged communities with groundwater contamination exacerbated by drought.
- Fifteen million dollars from the State general fund to address emergency water shortages due to drought.
- Thirteen million dollars from the general fund to augment the State and local conservation corps to expand water use efficiency and conservation activities and to reduce fuel loads to prevent catastrophic wildfires.
- Twenty-five point three million dollars from the general fund to be deployed to maximize the potential Federal drought assistance for providing food to those impacted by the drought.

- Twenty-one million dollars from the general fund for housing-related assistance for those impacted by the drought.
- One million dollars to continue the Save Our Water public education campaign.

On February 21, DWR sent letters to counties and well-drilling contractors asking for timely submission of well completion forms. This information will help DWR track increased use of groundwater and new well installation activities.

DWR and the Bureau continue to monitor water quality in the western, central, and southern Sacramento-San Joaquin Delta. The cross-channel gates along the Sacramento River near Walnut Grove were closed due to fishery concerns.

On February 12, DWR and the Bureau filed a petition with the State Board seeking authority to exchange water within the areas served by the Federal Central Valley Project and the State Water Project and vice versa.

On February 10, DWR announced the award of $153 million to help fund 138 separate water projects around the State, 35 of which will help communities cope with drought in the long-term.

DWR is working with the Bureau and the State Board to ensure an efficient process to transfer water between voluntary buyers and sellers. However, given the uncertainty of water supplies, few proposals for voluntary sales may be submitted.

On January 17, Governor Brown proclaimed a state of emergency and directed State officials to take all necessary actions to prepare for drought conditions. The Governor asked all Californians to reduce water consumption by 20 percent and referred residents and water agencies to the Save Our Water campaign for practical advice on how to do so. The Governor also directed State agencies to use less water and to hire more firefighters.

Key measures in the proclamation included:

1. Directing local water suppliers to immediately implement local water shortage contingency plans;
2. Ordering the State Water Resources Control Board to consider petitions for consolidation of places of use for the State Water Project and Central Valley Project, which could streamline water transfers and exchanges between water users;
3. Directing DWR and the State board to accelerate funding for projects that could break ground this year and enhance water supplies;
4. Ordering the State water board to put water rights holders across the State on notice that they may be directed to cease or reduce water diversions based on water shortages;
5. Asking the water board to consider modifying water quality control plan rules that require the release of water from reservoirs so that water may be conserved in reservoirs to protect cold water supplies for salmon and maintain water supplies;
6. And directing the State Department of Public Health to provide technical and financial assistance to communities at risk of running out of drinking water.

The Governor's proclamation is the fourth action taken by a Governor since 1987 to deal with drought on a statewide basis.

The Governor, through the emergency proclamation, directed his interagency drought task force to devise a plan to provide emergency food supplies, financial assistance, and unemployment services to communities hard-hit by drought-induced job losses.

AGRICULTURE-SPECIFIC ACTIONS

State officials are working closely with Federal agencies to provide assistance to farmers, ranchers and farmworkers in the most impacted communities. The California Department of Food and Agriculture has launched a one-stop Web site that provides updates on the drought and connects farmers to helpful State and Federal programs they can access during the drought. Farmers, ranchers and farmworkers can learn more at: *cdfa.ca.gov/drought/*.

- The site features links to crop insurance programs, crop disaster assistance, emergency farm loans and Federal water conservation program assistance.
- Governor Brown's partnership with the Obama administration on behalf of California has already led to millions of dollars in potential assistance for farmers and ranchers. Those opportunities are chronicled on the Web page.

The White House on Friday, February 14, 2014, announced emergency funding from several Federal programs to support drought response. This announcement was coordinated with President Obama's visit to Fresno County.

Emergency assistance includes:

- One hundred million dollars in expedited livestock disaster assistance to California farmers and ranchers. This funding, contained in the 2014 Farm bill, will be made available through the USDA in 60 days. Funding assistance can cover financial losses by California producers in 2012, 2013 and 2014.
- Sixty million dollars for California food banks to help families affected by the drought. Funding will be provided by the USDA's Emergency Food Assistance Program.
- Five million dollars of funding for conservation projects at California farms and ranches, provided by USDA's Environmental Quality Incentives Program.
- Five million dollars for emergency watershed improvements to enable activities such as stabilizing stream banks and replanting bare lands. Funds will come through the USDA's Emergency Watershed Protection Programs.
- Three million dollars for emergency grants to rural communities facing drinking water shortages. Funds come through USDA's Emergency Community Water Assistance program.
- Summer food programs: The USDA committed to expanding the number of Summer Food Service Program meal sites to 600 locations in drought stricken areas throughout the State.

LONG-TERM ACTIONS

There is broad agreement that the State's water management system is currently unable to satisfactorily meet both ecological and human needs, too exposed to wet and dry climate cycles and natural disasters, and inadequate to handle the additional pressures of future population growth and climate change. Solutions are complex and expensive, and they require the cooperation and sustained commitment of all Californians working together. To be sustainable, solutions must strike a balance between the need to provide for public health and safety (e.g., safe drinking water, clean rivers and beaches, flood protection), protect the environment, and support a stable California economy.

As we work on emergency actions to manage through this crisis, we are also taking proactive, long-term steps to prepare California for future droughts and flood. Our long term approach to preparing California's water future is captured in the California Water Action Plan which was released in January of this year. This plan will guide California's efforts to enhance water supply reliability, restore damaged and destroyed ecosystems, and improve the resilience of our infrastructure. We are working daily to balance needs and interests throughout the State on the overall long term sustainability of our water resources. This is not just about the current problem of this serious drought.

The California Water Action Plan has been developed to meet three broad objectives: more reliable water supplies, the restoration of important species and habitat, and a more resilient, sustainably managed water resources system that can better withstand inevitable and unforeseen pressures in the coming decades. Altogether, this plan centers on sustaining supplies of water for people, the environment, industry and agriculture.

This action plan lays out our challenges, our goals and decisive actions needed now to put California's water resources on a safer, more sustainable path. While this plan commits the State to moving forward, it also serves to recognize that State government cannot do this alone. Collaboration between Federal, State, local and tribal governments, in coordination with our partners in a wide range of industry, government and nongovernmental organizations is not only important—it is essential.

The Water Action Plan, over the next 5 years, will guide State efforts to enhance water supply reliability, restore damaged and destroyed ecosystems, and improve the resilience of our infrastructure.

With this plan, we recognize that water recycling, expanded water storage and groundwater management must all be part of the solution. We must also make investments in safe drinking water, restore wetlands and watersheds and make further progress on the Bay Delta Conservation Plan. All of these things are critical to the long-term solution.

Ten key actions identified:

- Make conservation a California way of life.
- Increase regional self-reliance and integrated water management across all levels of government.
- Achieve the co-equal goals for the Delta.
- Protect and restore important ecosystems.

- Manage and prepare for dry periods.
    - Expand water storage capacity and improve groundwater management.
- Provide safe water for all communities.
- Increase flood protection.
- Increase operational and regulatory efficiency.
- Identify sustainable and integrated financing opportunities.

There are many important components imbedded under each of these 10 actions. For the committee's benefit, let me highlight just two of these that go to the heart of this hearing's topic of addressing long-term solutions.

The Delta is California's major collection point for water, serving two-thirds of our State's population and providing irrigation water for millions of acres of farmland. We know too well the challenges of moving water through the Delta's fragile levee system with declining fish populations and historic restrictions on water deliveries. But, the status quo in the Delta is unacceptable and it would be irresponsible to wait for further degradation or a natural disaster before taking action.

While we are working to implement the Delta Plan, one component remains to be completed: The Bay-Delta Conservation Plan [BDCP]. State and Federal agencies will complete planning for this comprehensive conservation strategy aimed at protecting dozens of species of fish and wildlife in the Delta, while permitting the reliable operation of California's two biggest water delivery projects.

The BDCP will help secure California's water supply by building new water delivery infrastructure and operating the system to improve the ecological health of the Delta. It will also restore or protect approximately 145,000 acres of habitat to address the Delta's environmental challenges. The BDCP is made up of specific actions, called conservation measures, to improve the Delta ecosystem. It includes 22 conservation measures aimed at improving water operations, protecting water supplies and water quality, and restoring the Delta ecosystem within a stable regulatory framework. The project will be guided by 214 specific biological goals and objectives, improved science, and an adaptive management approach for operating the water conveyance facilities and implementing other conservation measures including habitat restoration and programs to address other stressors. As the Delta ecosystem improves in response to the implementation of the conservation measures, water operations will become more reliable, offering secure water supplies for 25 million Californians, an agricultural industry that feeds millions, and a thriving economy.

State and Federal agencies will complete the State and Federal environmental review documents; seek approval of the BDCP by the State and Federal fishery agencies; secure all permits required to implement the BDCP; finalize a financing plan; complete the design of BDCP facilities; and begin implementation of all conservation measures and mitigation measures, including construction of water conveyance improvements. Once the BDCP is permitted, it will become part of the Delta Plan.

We agree that we need to expand our State's storage capacity, whether surface or groundwater, whether big or small. We need more storage to deal with the effects of drought and climate change on water supplies for both human and ecosystem needs. Climate change will bring more frequent drought conditions and could reduce by half our largest natural storage system—the Sierra snowpack—as more precipitation falls as rain rather than snow, and as snow melts earlier and more rapidly. Moreover, we must better manage our groundwater basins to reverse alarming declines in groundwater levels. Continued declines in groundwater levels could lead to irreversible land subsidence, poor water quality, reduced surface flows, ecosystem impacts, and the permanent loss of capacity to store water as groundwater.

Among other actions to expand water storage, our plan supports funding partnerships for storage projects. The Brown administration will work with the Legislature to make funding available to share in the cost of storage projects if funding partners step forward. The State will facilitate among willing local partners and stakeholders the development of financeable, multi-benefit storage projects, including working with local partners to complete feasibility studies. For example, the Sites Project Joint Powers Agreement, formed by a group of local government entities in the Sacramento Valley, is a potential emerging partnership that can help Federal and State government determine the viability of a proposed off stream storage project—Sites Reservoir.

Over the next 5 years, this Water Action Plan will help us advance sustainable water management by providing a more reliable water supply for our farms and communities, restoring important wildlife habitat and species, and helping the State's water systems and environment become more resilient.

BAY DELTA CONSERVATION PLAN [BDCP]

The Draft Bay Delta Conservation Plan [BDCP] and associated EIR/EIS is now available for public review and comment. Lead State and Federal agencies recently extended the public comment period for the EIR/EIS by 60 days. The review period now totals 180 days stretching from December 13, 2013 to June 13, 2014. This extension will allow the public more time to review and comment on the public draft documents. The 180-day comment period is 4 times that of the required 45 days in order to ensure the public has plenty of time to review the draft documents. This extension is not anticipated to cause significant delays in the project, although it will likely extend the anticipated release date for the Final Draft BDCP and EIR/EIS. The public review draft documents are available online and electronically at libraries throughout the State serving as document repositories. DVD copies are also available on request.

As of February 12, DWR completed 12 public open house meetings throughout the State on the public review draft plan and associated EIR/EIS. More than 800 participants attended statewide. The meetings in Sacramento, Clarksburg and Stockton had the highest attendance, with Sacramento topping the list at 165 participants. A broad range of engaged stakeholder groups attended every meeting, including environmental, industry, business, water, and labor groups. Feedback has been that participants appreciated the format, and the ability to have one-on-one conversations with technical staff involved in the development of the project.

Our goal is to allow as many people as possible to provide comments, all of which will be carefully considered and will ultimately help shape the development of the final project.

If there's one thing these last 2 to 3 years demonstrate is that we need conveyance in place that can move water during wet years in a way that's safer for fish. Doing that allows us to lay-off in the dry years. We can't manage very easily through droughts without it. In the long-term, California must continue to focus on actions to modernize our water delivery system by completing the environmental planning process for the BDCP.

With the conveyance proposed in the Bay Delta Conservation Plan in place, the Central Valley this year would have an extra 800,000 acre-feet of water in the San Luis Reservoir. This effort to restore the Sacramento-San Joaquin Delta ecosystem and greatly enhance the water system's reliability is the best investment we can make right now in our water future.

In closing, it is important to note that California water policy moves in fits and starts tied to floods and droughts. When the rain and snow falls steadily and predictably, Californians tend to assume it will always be so. It's human nature.

We intend to take advantage of the public's hyper-focus on water issues this year to advance improvements to our water system.

East coast newspaper reporters lately have looked at our muddy reservoirs and declared that California has finally overreached and hit a wall.

We know better. We know it's because of our reservoirs—as well as our investments in water conservation, recycling, drip irrigation, groundwater recharge, and a host of other smart water management techniques—that we've been able to build a nearly $2 trillion economy in a State with hydrology that is as varied—both temporally and geographically—as California.

We are the most populous State, with the richest farm economy, and the most diverse natural heritage in the Nation. Our water system gets us through all but the most extreme, outlier years like this one without much sacrifice. We will cope, invest, and thrive.

On behalf of my colleagues at the State level, and our partners at the Federal level, I would like to thank you for holding this important hearing and providing this opportunity to provide testimony. Thank you for your attention to these issues.

The CHAIRMAN. Thank you very much for your testimony.

Next I will recognize Mr. Steve Knell, P.E., the General Manager of the Oakdale Irrigation District.

Mr. Knell, you are recognized for 5 minutes.

STATEMENT OF STEVE KNELL, P.E., GENERAL MANAGER, OAKDALE IRRIGATION DISTRICT, OAKDALE, CALIFORNIA

Mr. KNELL. Thank you, Mr. Chairman, committee members, for taking the time to come to California to hear the concerns and issues surrounding what is now the State's 3rd year of a drought. My presentation will cover the drought impact——

The CHAIRMAN. Move the microphone closer to you, if you would.

Mr. KNELL. Is this better?

The CHAIRMAN. That is much better. Thank you.

Mr. KNELL. Thank you. My presentation will cover the drought's current impacts on the Oakdale Irrigation District and offer a suggested action that should be considered to address both the immediate and long-term needs of water supply and reliability in our State.

Compared to many districts, OID has been less affected by this 3rd year of drought. I know now today we are the ant, as spoken about earlier. We have prepared for this event by investing significant amounts of time and resources into making our district better.

Since 2006, we have put in $50 million of modernization technologies into our district to advance our district so that we are better delivering water to farms. When we deliver water better to farms, farmers can better manage their water on the farm, making for a rollup of greater efficiencies within irrigation districts. These technologies are funded solely by water transfers of our conserved water to agricultural and municipal water districts in the State.

While OID is not immune to the effects of this 3rd year of drought, the options gained through these investments soften those impacts.

As a State, California has been less successful at developing a drought plan. We choose to spend efforts on the Lower San Joaquin River Restoration Program to achieve 500 returning spring-run salmon. We have gravitated to choosing to build bullet trains over water storage or water conveyance. The State believes it can restore a delta to its once pristine condition simply by taking 35 percent of the unimpaired runoff from our watersheds and sending it out to the ocean. We have taken 10,000 acres of the most productive farmland in the world out of production at a time when a significant portion of Americans still go to bed hungry. The failure to prioritize, plan, and invest leaves California ill-prepared and ill-equipped to address the human and financial consequences brought on by this drought.

Water managers in California support increased storage. Currently, 47 percent of the State's available water is dedicated to environmental purposes. Agriculture takes 42 percent, and the urban part of the equation gets 11. More storage makes no sense if that 47 percent share of environmental water escalates under the vision of the State's Water Resources Control Board. Storage only makes sense if you have water to put in storage.

New Melones is a large federally owned storage reservoir on the Stanislaus River. It has the capacity to hold 2.4 million acre feet, while the annual yield of the basin is 1.1 million acre feet, a very smartly built dam. Unfortunately, the ability of the United States Bureau of Reclamation to store water in New Melones is restricted by a Biological Opinion which over-commits the reservoir's water to downstream uses, leaving New Melones with underutilized storage each and every year.

OID and our sister districts, South San Joaquin Irrigation District, divert water from the Stanislaus River but do not have the right to store water in New Melones without Federal permission under a Warren Act contract. Over the past several dry years, the need to utilize storage in New Melones has become increasingly evident. In 2010, OID and SSJID nearly lost the ability to move 80,000 acre feet to the San Luis Delta-Mendota Water Authority. There was difficulty in matching up the timing of water availability to pump access at the Federal pumps in the delta. Through extraordinary cooperation, the water eventually moved.

In 2012, that ability was not there, and that water did not move to the west side. If storage in New Melones had been available, we could have changed that.

Our initial review indicates somewhere between 20,000 and 40,000 acre feet could be put into storage annually behind New Melones. The cost to raise Shasta Lake to get 20,000 to 72,000 acre feet is $280 million to $360 million. The cost to raise the San Luis Reservoir is $360 million to gain 130,000 acre feet. To store 20,000 to 40,000 acre feet annually in New Melones would not cost a dollar. In fact, the Federal Treasury could actually gain $1 million a year in storage fees. Imagine that.

Access to storage in New Melones provides multiple benefits, as is provided in my written testimony, chief of which is the reliability to meet exporter needs in water years in which storage over-availability is a priority. Congressman Denham introduced H.R. 2554 to address the issue of New Melones storage. This bill was approved by the House and included in a final compromise bill that was sent to the Senate. We hope that the Senate will act expeditiously in the passage of this legislation.

And I wish to thank the committee for hearing the views of the Oakdale Irrigation District.

[Applause.]

[The prepared statement of Mr. Knell follows:]

PREPARED STATEMENT OF STEVE KNELL, P.E., GENERAL MANAGER, OAKDALE IRRIGATION DISTRICT, OAKDALE, CALIFORNIA

OPENING REMARKS

I would like to thank the Natural Resources Chairman and committee members for taking the time to come to California and hear the concerns and issues surrounding what now is California's 3rd year of drought.

My presentation will cover the drought's current *impacts* on the Oakdale Irrigation District [OID] and offer a suggested action that should be considered to address both the *immediate* and long-term needs of water supply and water reliability.

IMPACTS OF THE DROUGHT

Compared to many districts, OID has been less affected by this 3rd year of drought. OID's preparedness is a result of the investment of incredible time and resources preparing for this event. OID has planned, financed and implemented modernization technologies that have allowed itself to be more efficient at delivering water to its farmers. The benefits from that effort and the investments made are visible in times of drought. While OID is not immune to the effects of this third year of drought, the options gained through its investments soften those impacts. OID began its modernization and infrastructure replacement program in 2006; its projected costs were $168 million back then. OID has spent $50 million to date and still has a way to go. That planning vision began back in 2003 when the OID Board of Directors voted for OID to embark upon this course.

As a State, California has been less successful at developing a focused plan and investing time and resources to achieve drought preparedness. We are choosing to spend billions of dollars on the lower San Joaquin River Restoration Program to achieve 500 returning Spring Run Salmon, when communities in our valley do not have safe, affordable and reliable drinking water. We have gravitated to choosing to build bullet trains over water storage or water conveyance facilities. The State of California believes we can ''restore a delta'' to its once-pristine condition simply by taking 35 percent of the unimpaired runoff from our watersheds and sending it out to the ocean. We have taken tens of thousands of acres of the most productive farmland in the world out of production when 25 percent of the American population goes to bed hungry every night. The failure to prioritize, invest and plan leaves California ill prepared and ill equipped to address the human and financial consequences brought on by this third year of drought.

IMMEDIATE AND LONG TERM SOLUTIONS

More Reservoir Storage

Water managers in California support increased storage; whether for urban or agricultural use. However, water storage projects require significant investment, planning, and long-term reliability. Storage projects are useless if the State of California, through the State Water Resources Control Board, is focused upon taking more water out of existing storage and sending it to the ocean.

Currently, 47 percent of the State's available water is dedicated to environmental purposes. Agriculture takes 42 percent and the urban part of the equation gets 11 percent. More storage makes no sense if that 47 percent share of environmental water escalates, as is the current vision of California's Department of Water Resources.

While agriculture has made great strides in conservation, like OID, we can do more if there is an incentive to build and pay for projects to conserve water. In OID's situation, the ability to pay for more expensive conservation is diminished because OID has limited ability to store and use conserved water. Storage only makes sense if you can put water into storage.

Maximizing the Use of Existing Storage

New Melones is a large federally owned storage reservoir on the Stanislaus River. It has the capacity to hold 2.4 million acre feet, while the annual yield of the Stanislaus River basin is 1.1 million acre feet. The ability of the U.S. Bureau of Reclamation to store water in New Melones is restricted by a Biological Opinion [BO]. That BO calls for such large in-stream flow releases annually, that over time the resource becomes over-committed. Due, in part, to the BO requirements, New Melones has a significant amount of storage capacity that is not used each and every year.

OID and SSJID divert water from the Stanislaus River, but do not have a right to store water in New Melones without Federal permission through a Warren Act Contract. OID and SSJID have smaller storage facilities upstream of New Melones. Despite OID's limited ability to store water, it has invested in water use efficiency. In 2001, it took OID 255,000 acre feet to meet its crop water demands on-farm. Twelve years later and $50 million in conservation investments, it now takes OID 235,000 acre feet to do the same job. OID markets the surplus water generated from its conservation efforts to finance and fund its modernization programs.

Over the past several years, the need to utilize storage in New Melones has become increasingly evident. For example, in 2010, OID and SSJID agreed to transfer water to the San Luis Delta-Mendota Water Authority [SLDMWA]. Pumping restrictions and project operations almost defeated this transfer because it was difficult to match up the timing of water availability and the ability to pump. If storage in New Melones had been available, that difficulty could have been avoided. The same was true in 2012 when OID and SSJID discussed moving supplemental supplies to SLDMWA, only to be told there was no capacity at the pumps. Since there was no capacity, and the Districts could not store the water in New Melones, this transfer did not occur. A lost opportunity in a dry year.

OID and SSJID have analyzed the historic hydrology of the Stanislaus River basin and the current operations under the BO, and have found time periods of up to 35 years in length where additional water could have been put into New Melones without spill. As part of its planning efforts, OID has looked at storing its conserved water, coordinated river releases to meet regional regulatory obligations on the Merced, Tuolumne, and Stanislaus Rivers, and water exchanges with MID/TID as mechanisms available to generate water for storage. Our initial review indicates somewhere between 20,000–40,000 acre feet could be put into storage annually. The costs to raise Shasta Lake and get 20,000–72,000 acre feet of water is $280–$360 million. The costs to raise the San Luis Reservoir is $360 million for 130,000 acre feet of water. To store 20,000–40,000 acre feet annually in New Melones would cost

$0. In fact, the Federal Treasury would receive up to $1,000,000 in Warren Act contract payments.

Access to storage in New Melones provides multiple benefits;

1. Increased stored water without any additional capital costs, construction, or major environmental permitting.
2. More storage under a Warren Act Contract means more revenue to the Federal Government via storage fees. OID has even offered to pay the Federal Government in ''water'' as opposed to ''money'' to benefit their purposes.
3. OID's water transfers and the release of that water could be coordinated on a fish-friendly flow schedule that benefit the environment.
4. Storage in New Melones would afford OID and SSJID the ability to move water when needed, as needed on an annual basis to better meet the needs of all exporters, both on the Federal and State Water Project, when pump capacities are an issue.
5. Storage in New Melones would afford greater reliability to those same exporters as a carryover option in water years that *storage* over *availability* is a priority.
6. Storage in New Melones could be enhanced by inter-basin, eastside transfers and exchanges.

CONCLUSION

Congressman Denham introduced H.R. 2554 to address the issue of New Melones storage. This bill was approved by the House and included in a final compromise bill that was sent to the Senate. We ask that the Senate act expeditiously in the passage of this legislation.

I wish to thank the committee again for their time and for listening to the views of the Oakdale Irrigation District.

The CHAIRMAN. Thank you very much, Mr. Knell.

And last, and certainly not least, we have Mr. Kole Upton from Chowchilla, California.

Mr. Upton, you are recognized for 5 minutes.

STATEMENT OF KOLE UPTON, CHOWCHILLA, CALIFORNIA

Mr. UPTON. Thank you, Mr. Chairman and the committee. I am a family farmer. I live on my farm in Merced County. I farm with my brother and two sons. The farm was started in 1946 when my dad returned from 3 years in Europe in World War II. I started farming in 1971 after I spent 6 years in the Air Force.

We get our water from the underground aquifer, but also from Friant Dam and an auxiliary dam in Buchanan Dam. We are facing the same situation that the valley faced in the 1920s and the 1930s, and that was when the underground aquifer was being depleted, and that is why the government decided to build Friant Dam, so they could bring surface water into the Friant service area, which is about a million acres on the east side. And the people that came out and developed this land were people from the Depression and World War II, and the government passed the reclamation law, which gave people an opportunity to build farms, support their families and build these wonderful communities we have all the way from Merced to Bakersfield.

But this whole system, this whole society is under attack now, and it is in jeopardy because of the environmental laws that we have, and we shouldn't be blaming the judges because more than one judge has said, hey, don't blame me; if you want to change, then change the law so they can interpret it and so the agencies can interpret them with common sense.

The first item I want to talk about is the San Joaquin River Settlement. I was involved in that. I was one of the negotiators, and the goals were laudable. Senator Feinstein and Congressman Radanovich asked NRDC and Friant to try to come to two goals. Number-one goal was to try to get a self-sustaining salmon fishery back on the San Joaquin River. The number-two and co-equal goal was for the farmers and the east side communities to get their water back. Those were the rules.

So we did that. It was signed on September 13, 2006. In addition, Senator Feinstein made us all sign a blood oath that we would, in good faith and integrity, try to achieve both those goals. Well, shortly thereafter and before it was signed in 2008 by President Obama, NRDC and the other environmental organizations started chipping away by getting themselves involved in court cases and agencies' decisions and that kind of thing.

The bottom line was that first they were able to successfully kick out a lot of water that was going to the west side. But in addition, they knocked out our ability to recirculate our water under the water management goal. The key to that was taking the water down to the southern Friant districts. Well, we couldn't do that anymore because of these environmental actions, and yet we can't get anybody from the Senate or anybody else to hold NRDC accountable to this. The lady that is not here today, she actually came from NRDC 3 or 4 years ago, so I am not expecting a whole lot of help from her in this situation. So we are going to need—we need help from the Congress to change some of these laws.

In addition to not getting the water management goal done, the fish are not going to make it on the San Joaquin. NRDC's own data shows that it is going to be too hot because of climate warming, whatever you want to call it. It is not going to work, so they can't have it both ways.

And the last thing which I think Mr. Murillo has told me before, they are short of money. They don't have enough money to finish this thing. So they are releasing water down the river when it is available for 10 projects, and none have been done, and the money is not available to finish it.

In addition, the third part of this agreement was no harm, no foul to third parties. Well, I can tell you that some of the third parties along the San Joaquin River where they release these restoration flows, farmers have water coming up in the root zones of the permanent crops, and they are hurting them.

So this thing is a disaster. It needs to be changed. Your bill addresses this.

The other thing is this groundwater subsidence that is occurring because we are using so much underground. You have to love the environmentalists. Now they are saying, we have to come in here and we have to fix this groundwater problem, these darn farmers are pumping too much from the underground. What do they expect? They have taken all our surface water. What the hell are we going to do. That is the only choice we have, all right, is the underground.

[Applause.]

Mr. UPTON. OK. We need three things from you folks. Number one is we need environmental accountability. Right now, the urban user, the business user, the farm user, we are all required to conserve and be accountable. Hey, that is good. I understand that. Environmental releases, there is no accountability. There is no accountability. Millions of acre feet have gone out, and the purpose for which it was intended has not helped anybody. So we need that to have the same accountability. If it is not working for the fish or whatever, then bring it back to the people that are already using it and leave it there.

The second thing is we need to re-do the River Settlement.

The third thing we need is Temperance Flat. There is no sin in investing in a dam for the future food supply of the people of the United States. You have to have a storage so we can do water banking. People say water banking is the answer. It is one of the answers, but you have to have two things. There is too much water coming down the river for it to be able to percolate in a water bank.

I will close now by saying that if we don't do this, then we are going to have to fallow hundreds of thousands of acres, like Congressman Nunes said. And I have to tell you, I think we have an obligation to people that live here, that have taken the sacrifice of our forefathers in building this area and changing it from a desert into a garden, to fight for what they thought was right. I spent 6 years in the military. My dad spent 3 years in the military. I have to tell you, I feel like my own farm, my own family is under an attack by my own government.

[Applause.]

Mr. UPTON. Last, I would say the message I would have is I commend you folks, but I would give a message to the Senate: the time for talk is over. It is time for them to do something.

[Applause.]

[The prepared statement of Mr. Upton follows:]

PREPARED STATEMENT OF KOLE UPTON, CHOWCHILLA, CALIFORNIA

Mr. Chairman and members of the committee, it is an honor and privilege to appear before the House Committee on Natural Resources. I appreciate the opportunity to testify concerning the subject of the hearing: "California Water Crisis and Its Impacts: The Need for Immediate and Long-Term Solutions".

We are family farmers who live on our farm. It was started when our father returned from World War II. With my brother and sons, I grow pistachios, almonds, wheat, corn, barley, oats, and occasionally pima cotton. As most framers in the Friant service area, we are in a conjunctive water use area. Our water comes from both surface water supplies and the underground aquifer.

I am appearing as an individual at this hearing and not as a representative of any of the water or agricultural organizations of which I am board member.

THE PROBLEM

This latest drought has magnified and exposed the water crisis being inflicted on the east side of the San Joaquin Valley. The availability of adequate and affordable surface water is essential to the future of this valley. It was the depletion of the underground aquifers in the 1920s and 1930s that led to the building of Friant Dam.

Remarkably, we are facing the same scenario now. However, it is not because of the lack of surface water, it is due to the surface water being "reallocated" because of the San Joaquin River Settlement and the enacting legislation. Surface water that should have been used in lieu of underground water and used to replenish the underground aquifers has instead been redirected to flow to the ocean. Had the redirection of this immense amount of surface water resulted in some magnificent environmental achievement or the saving of some species, then perhaps it might have been worth it.

However, the reallocation from east side users has not resulted in any environmental improvements. The San Joaquin Restoration water releases have been totally wasted because none of the projects to get the River ready for salmon have been completed. With no other options, farmers have turned to the underground aquifers to sustain their crops. Now, we are in crisis with the underground being depleted at an unsustainable rate.

The proposed solution by the environmental community and its allies in the government is to demand regulation of underground pumping. They allege farmers are acting irresponsibly by depleting the underground, and are trying to use this ruse as a reason to further hamstring and control water usage by farmers.

THE SAN JOAQUIN RIVER SETTLEMENT

To understand this situation, we must first review the San Joaquin River Settlement. The concept of settlement was advanced by Senator Feinstein and Congressman Radanovich. There were two co-equal goals: 1. Attempt to revive a self-sustaining salmon fishery on the main stem of the San Joaquin River (Restoration Goal); and 2. Mitigate the water losses of the folks that have depended on this surface water for decades (Water Management Goal).

The Settlement was signed on Sept. 13, 2006, and Senator Feinstein required all parties to sign a "blood oath" promising to abide by its terms, conditions, and goals. The key to the Water Management Goal was the ability to recirculate the restoration water back to the southern Friant districts once it reached the delta. However, the Natural Resources Defense Council [NRDC] the primary environmental plaintiff in the Settlement and the negotiations aggressively continued to participate and inject itself into critical litigation and regulatory matters after the signing of the Settlement with a view of doing damage to delta conveyance and thus to recirculation efforts.

The bottom line is that the water losses cannot be mitigated and the co-equal Water Management Goal is a sham.

In addition, the goal of a self-sustaining salmon fishery is also not achievable. First the funding has dried up, and none of the projects required to get the river ready for salmon have been completed. Nevertheless, the environmentalists and some government officials continue to demand that hundreds of thousand of acre-feet be released to the ocean anyway. Also, the environmentalists own data shows that the water temperatures caused by global warming will be too hot for salmon to survive.

Finally, the promise of no harmful impacts to third parties as a result of actions involving the Settlement has also been broken. An example is the farmers along the San Joaquin River who have had their permanent crops damaged by water seeping up in to the root zone because of restoration flows.

THE IMMEDIATE SOLUTION

For the first time in many years, there is proposed legislation in both the House and the Senate to address the water situation in California. For the east side of the San Joaquin Valley, it is imperative that the revision of the San Joaquin River Settlement be "on the table" and part of the legislation. The revision is simple. Change the River Restoration goal from a self-sustaining salmon fishery to an extension of the current 40 mile, robust fishery that currently exists below Friant Dam. This concept will provide us with a live fishery 360 days/year and allow achievement of the Water management Goal.

In addition, it will save billions of dollars. Some of these savings could be used to enhance the salmon fisheries currently in existence that are in the cooler climates required for salmon viability. Harmful Third party impacts will also be eliminated. The result would be a live river, more total salmon, and the return of the availability of hundreds of thousands of acre-feet of surface water that is essential to the east side.

LONG TERM SOLUTIONS

There are several long term solutions required for this area to be able to maintain its ability to feed the Nation and the world. They are:

1. An appropriate revision of the Endangered Species Act [ESA] that allows for the worthwhile goals of the act to be achieved without decimating areas like the San Joaquin Valley, and the food supply of the United States. Concurrent

with that legislation, a proposed law requiring environmental water releases be held to the same standards for efficiency and accountability as required of urban and agricultural uses. Water is a public resource and should not be wasted by any user. So, if an environmental water release is not accomplishing the task for which it is being released, then it should be made available to the other water users so it may be beneficially used for society.

2. A water balance analysis is done for the San Joaquin Valley so that residents and decisionmakers know the extent and seriousness of the situation. Following that analysis, a determination be made as to how many acres of productive farm land must be permanently fallowed to get the area in to water balance.
3. A new dam built at Temperance Flat with public funds. Twice in the last 20 years, flood events have resulted in the loss of millions of acre-feet of water because Friant Dam is too small. Water banking by itself cannot address this problem because it takes time for the water to percolate in water banks. The additional storage will provide the time to store the water and then be released over time. In addition, this water can be available for in-lieu recharge which is the most effective means of underground replenishment.

There is nothing sinful about a society investing in its own infrastructure. A new dam is the investment in the future food security of the United State. It would provide the additional water needed to help restore some water balance to the area, as well as significant flood control benefits. However, without revising the San Joaquin River Settlement, a new dam would be virtually useless. The only solution would then be to permanently fallow hundreds of thousands of more acres of productive farm land.

CONCLUSION

We are at a crossroads for the east side of the San Joaquin Valley. For some 50 years, we have thrived due to the foresight, planning, and wisdom of our forefathers. Leaders of both political parties worked together to provide an opportunity for the World War II generation by building Friant Dam and enacting Reclamation Law. This generation responded magnificently by creating a robust society of small and medium sized communities embedded in 1,000,000 acres of productive farm land. This land has nourished this country and the world. It has been a government program that worked.

Now, all that is at risk not because of any continuing natural calamity, but because of a continuing series of overreaching environmental laws passed by Federal and State legislators and enforced by bureaucrats and judges. The only solution is to revise these onerous laws. The time to act is now.

The CHAIRMAN. Thank you very much, Mr. Upton.

I also want to say that at the entrance here there are forms if any of you would like to make a comment as part of the record for our committee hearing, or you can go to our Web site, which is *naturalresources.house.gov,* and go to ''Contact,'' and you can put your comments in there if you would like, and we obviously invite your comments.

I want to thank all the witnesses for your testimony. Now we will start the process here where we Members will ask questions of some of the witnesses. I will start with myself.

Mr. Murillo, I want to ask you a question. I will ask my staff if they can put up a graph because this graph, I think, at least from my perspective, not living here all of the time, certainly tells a lot. It is a graph going back from 1952 to 2013. The blue line represents the storage in millions of acre feet from the Central Valley Project starting in 1952 all the way up to 2013. You can see that blue line has been fairly constant, with the exception probably of some drought areas, 1976 probably being the most obvious one. I remember that was a big drought in my State of Washington.

The red square is the initial Ag Service allocation to agriculture. Since we are talking about agriculture, that is what the red square was. The green triangle was the final Ag Service allocation. Now, if you look at that graph starting from 1952 through 1990 or maybe 1991, with the exception of 1976 when that really was very much a drought, the allocation, the initial allocation and the final allocation was sufficient. It was always higher than what the storage was.

But all of that changed, it appears, in 1992, where you can see that the initial Service allocation and the final allocation was all over the board, mainly during that time period under what the CVP storage was.

So, Mr. Murillo, I just wonder if you can explain to me, somebody from outside the State, what happened in the early 1990s to change this graph when the one constant all the way through, at least from the graph, was that the storage was constant. So if you could explain that to me, I would very much appreciate it.

Mr. MURILLO. Thank you for the question, I appreciate it. So the CVPIA, one of the requirements of the CVPIA is that we set aside CVP yield, water yield, about 800,000 acre feet a year. We provide that for in-stream flows. So I think part of what you are seeing is you are seeing some of the requirements that have been imposed on us by the CVPIA.

The CHAIRMAN. The CVPIA, explain what that was again.

Mr. MURILLO. The Central Valley Project——

The CHAIRMAN. Improvement Act? Is that what it was?

Mr. MURILLO [continuing]. Improvement Act.

The CHAIRMAN. OK. And that is legislation——

Mr. MURILLO. It was legislation——

The CHAIRMAN. I was not in Congress at the time. That is legislation that was passed in the early 1990s through the Congress.

Mr. MURILLO. Yes, it was, and it was for us to take into consideration equally fish and wildlife demands. So we were supposed to put those equal to power generation, Ag delivery. So we were supposed to hold it at a high priority. So what that did is it required us to provide additional flows, in-stream flows to meet some of those concerns.

I think what you are seeing here, at least part of the impact you are seeing here is some of that 800,000 acre feet that is being allocated to meet the environmental requirements of the CVPIA.

The CHAIRMAN. Well, Mr. Murillo, I have to say, toward the end, when I look at these red dots where you have—again, comparing this with storage, comparing it to the final allocation, the initial allocation—final allocation, the storage seems to be pretty much constant. But the allocation, both initial and the final, seems to be much lower. Maybe I am missing something. You said 800,000 acre feet. Why would it be lower in the out years here? What is the reason for that?

Mr. MURILLO. Well, like I said, I think part of the reduction and probably the allocation and the final allocation that we present is there are a number of factors that we have to take into consideration these last several years. So CVPIA is one. That is something that we do have to take a look at. We talked about this before.

There are just a number of other requirements. We have the water quality requirements that we have to continue to meet.

The CHAIRMAN. Mr. Murillo, my time is rapidly coming out. Again, I ask that question as somebody not from the State. When I see this graph here, I see a dramatic change because, obviously, of legislation that happened. I think there needs to be an explanation, and maybe some of my colleagues who live with this better than I can follow up. So thank you for that.

I will yield back my time and recognize Mr. Costa.

Mr. COSTA. Thank you very much, Mr. Chairman.

Larry Starrh and Mark Watte and a few of our folks here, we have worked together over the years, and with your family, and I think a number of your comments clearly indicate it. I mean, we are kind of preaching to the choir here.

My question to maybe you, Mark, and Larry, and if some of you want to opine in, we have 38 million people in this State, as we noted, and we have a broken water system, as we have noted, but we have 4 million people that live here in the valley, and if it were just us to make the determination, obviously, I think it would have been done a long time ago.

Having said that, what do you think is going to be necessary to get people in southern California, people in the Santa Clara Valley that, by the way, get water out of San Luis, people in the delta to understand that their water is at risk as well? Because 4 million people by ourselves cannot convince 38 million people in the State by ourselves that we have a broken water system.

Mr. STARRH. Well, I will tell you, Benjamin Franklin said it. He said, "When the well is dry, we know the worth of water."

Mr. COSTA. I know. But southern California doesn't believe that water is at risk.

Mr. STARRH. If I was really good at going out and being a publicist and a marketer of the problem, I think until you have a real tragedy—and I hate to say that—either the price of food has to come——

Mr. COSTA. And maybe the silver lining in this crisis, if there is one, is that the price of food will get to a point that maybe people will wake up. I don't know.

Mark, do you have a thought?

Mr. WATTE. Well, the common theme through the whole day today is the issues we have with the Endangered Species Act. That law is just out of control.

Mr. COSTA. I agree.

Mr. WATTE. And I think until our elected officials will face that fact and push back against some of the most extreme——

Mr. COSTA. Yes, but we respond to our constituencies. And if our constituencies in southern California don't think they have a problem——

Mr. WATTE. Well, I think they are listening to a really tiny, small constituency. The bulk of the constituencies are disengaged. They are not aware.

Mr. COSTA. Oh, I would agree with that point, too.

Mr. WATTE. So I think it is up to the elected to begin the process——

Mr. COSTA. In Santa Clara, their intake is 30 feet above San Luis Reservoir, and they are going along like there is not a problem.

Mr. WATTE. Well, I just think some leadership from the elected is a beginning point.

[Applause.]

Mr. COSTA. But you have, Mark, you have leadership here, but we don't represent southern California. You know, I mean, let's——

Mr. WATTE. I understand that, I understand that.

[Voice.]

Mr. COSTA. Well, it is right here. I voted for all of those bills.

[Voice.]

Mr. COSTA. Excuse me. I have another question.

The CHAIRMAN. Mr. Costa has the time, and please respect the fact that he has the time.

Mr. Costa, you are recognized.

Mr. COSTA. Thank you very much.

Ms. Beland, you spoke about what is currently going on. But does the Governor understand that we have a broken water system?

Ms. BELAND. I would say yes, we very much understand that fact, and we have——

Mr. COSTA. And the proposal to fix this broken water system by the Governor is what?

Ms. BELAND. The Bay Delta Conservation Plan.

Mr. COSTA. And that does what? We don't want to get into—people just want to know. Does that include storage? Does that include fixing the delta? Does that include creating flexibility in how the State and Federal pumps are operated?

Ms. BELAND. The BDPC, compliant with the California Water Action Plan, would do all of those things. It would allow us to fix our conveyance system——

Mr. COSTA. By building two tunnels.

Ms. BELAND [continuing]. By building two tunnels——

Mr. COSTA. Additional storage?

Ms. BELAND [continuing]. So that we can reliably move water to storage, and——

Mr. COSTA. Have you sited the storage? Are we talking about Shasta sites? Are we talking about San Luis? Are we talking about Temperance Flat?

Ms. BELAND. We have not sited storage. We are looking for storage partnerships because the State——

Mr. COSTA. Well, there has to be a cost sharing. I voted for the Valadao legislation. I was proud of that fact. But we don't provide any Federal funding in that legislation, which is why I introduced the other legislation, and that was understanding the issues that Mr. McClintock raised. I am more than happy to work with you on that point.

Mr. Murillo, quickly, we talked about the storage, and you talked about the completion dates of the studies. When could we begin construction if we were able to reach an agreement between the State and locals on cost sharing on the construction of each of these individual projects? Let's take Shasta, for example. When could we begin there?

Mr. MURILLO. Actual construction itself? That all depends on, once again, once we complete the Environmental Impact Statements and——

Mr. COSTA. Well, but, I mean, it has gone way too long. Mike Connor acknowledges that; you acknowledge that. I mean, when could we begin construction? You are going to have to answer those questions.

My time has run out. But I just think that until we can tell people up here definitively, if we reach a cost-sharing arrangement, that we can begin construction next year or in 18 months, nobody is believing us.

Mr. MURILLO. Well, like I said, the final feasibility study and the final EIS for the Shasta will be completed this year. If we have cost-share partners, we should be able to get moving——

The CHAIRMAN. The time of the gentleman has expired.

The Chair recognizes the gentleman from California, Mr. McClintock.

Mr. MCCLINTOCK. Just to quickly pick up on that point, Shasta was designed at an 800-foot level. It was built to 600 because we didn't need the capacity then. That additional 200 feet would mean 9 million acre feet of additional storage. How much are you actually proposing to increase Shasta by?

Mr. MURILLO. What we are looking at is an 18.5-foot raise.

Mr. MCCLINTOCK. Eighteen-and-a-half feet when it was designed to be raised another 200 feet.

Now, you say that the feasibility study will be done then. The feasibility study is to meet all of the existing laws and regulations and edicts, which tend to make these completely infeasible. So isn't this just a shell game?

That is a rhetorical question.

[Laughter.]

[Applause.]

Mr. MCCLINTOCK. Ms. Beland, the Burns Porter Act produced the entire State Water Project. That was 21 dams, 7 million acre feet of water storage, 3,000 mega-watts of hydro-electricity, 700 miles of canals. If you do the inflation adjustment for the California Water Project, it is about $17 billion in today's money. The Governor now proposes $14 billion for the cross-delta facility that adds zero additional water storage and zero additional hydro-electricity.

Wouldn't that $14 billion be better spent building additional capacity? For example, $6 billion to raise Shasta to its full design elevation of 800 feet, build the Auburn Dam, which was half completed in the 1970s. That by itself, those two projects would more than double storage on the entire Sacramento system.

Ms. BELAND. Well, the Bay Delta Conservation Plan is only a portion of what the Governor is proposing that we do in the State. Our State Water Action Plan includes storage, both above ground and below ground.

Mr. MCCLINTOCK. How much storage? Anywhere close to the 9 million acre feet in completing Shasta, or the 2.3 million acre feet of completing Auburn?

Ms. BELAND. It is going to rely and depend on those partnerships that can step forward. We don't have the financing to build——

Mr. MCCLINTOCK. I think the obvious answer is no.

Mr. Watte, are we ever going to solve our water shortage without fundamental reform of the Endangered Species Act?

Mr. WATTE. Sorry?

Mr. MCCLINTOCK. Are we ever going to solve our water shortage without fundamentally reforming the Endangered Species Act?

Mr. WATTE. Absolutely not.

Mr. MCCLINTOCK. Ms. Beland, Mr. Starrh referenced 800,000 acre feet of water that was released out of Oroville Shasta in Folsom Dam this fall. Those of us in Sacramento watched the Sacramento River at full flood, wondering what in the world were you people thinking. Our subcommittee has requested the release orders, many of which are completely without explanation.

Let me put it to you: why was this water released last fall knowing full well that we were heading into a potentially catastrophic drought, draining the Folsom Lake almost completely of its water?

Ms. BELAND. We weren't anticipating the drought to continue in——

[Voices.]

Mr. MCCLINTOCK. Thank you for your candor.

Mr. Upton——

[Voices.]

The CHAIRMAN. Please come to order so we can continue.

Mr. MCCLINTOCK. Mr. Upton, you mentioned the incestuous relationship between extremist groups like the NRDC and policymakers in State government, particularly the State Water Resources Control Board. How widespread is this relationship?

Mr. UPTON. Well, I don't know. But I know that in a lot of the agencies and when you go to talk to people, you are facing the same people that you faced years ago when you were dealing with environmental organizations.

Mr. MCCLINTOCK. So you would say that relationship is pretty extensive now throughout the State government?

Mr. UPTON. Yes, it is very extensive, and it is very extensive on a lot of the staffs, especially in Sacramento and in DC. People come out of the environmental organizations and get on the staffs.

Mr. MCCLINTOCK. And it is also true in the Federal Government?

Mr. UPTON. Yes, also true of the Federal Government. It is very difficult to explain to somebody that doesn't understand Ag, doesn't understand water, and doesn't appreciate it. So before you can even convince them of anything, you first have to educate them, and a lot of times they don't want to be educated.

Mr. MCCLINTOCK. Mr. Knell, we haven't had a major reservoir since New Melones in 1979. You mentioned the need to build more capacity, but also to more carefully use our existing capacity. You pointed out that much of that capacity has been squandered to meet various governmental environmental mandates. Do you have any idea how much of our existing capacity—forget building new storage. Just with our existing capacity, how much of that is being squandered due to these decisions and laws and regulations?

Mr. KNELL. Well, I can't speak for the State. I can speak for Melones. I mean, the reservoir is——

The CHAIRMAN. Please speak into the microphone. Pull it a little closer.

Mr. KNELL. I am sorry. Melones is, speaking from our basis, Melones is over-committed to meet downstream resource demands for fisheries, water quality, and all the rest.

Mr. MCCLINTOCK. Can you just give us a percentage? How much of New Melones is being used for people, and how much of it is being used for fish?

Mr. KNELL. Is it 60/40? It is 60/40.

The CHAIRMAN. The time of the gentleman has expired.

The Chair recognizes the gentle lady from Wyoming, Mrs. Lummis.

Mrs. LUMMIS. Thank you, Mr. Chairman.

A question for anybody on the panel. Can anyone tell me whether these flows that have cutoff irrigation water and are intended to help the recovery of the delta smelt have indeed done so? Is the delta smelt survivability increasing because of these flows? Can anyone answer that verbally?

Mr. Delgado?

Mr. DELGADO. Yes. As far as the flows, there is water coming in from northern California. It is from the Sacramento River. The delta is a big mess right now. It is a cesspool of pollution. And basically, in my opinion—I might be wrong, but the water that has been legally stolen from us has been diverted through Sacramento River into the delta to clean up all the sewage from the cities along the coast, and then thrown basically back into the San Francisco Bay area to clean up the pollution from San Francisco and clean up the pollution from the oil refineries. They are using the excuses about the delta smelt and the fish, but they don't even exist anyway.

The biggest problem we have is pumping restrictions. The last rain we had, we had rain, and you wouldn't believe—they would startup the pumps to restore and recharge the San Luis Reservoir. Instead, the pumps were only pumping like 10 percent of capacity, and the rest of the water basically went out into the ocean. That is the way I look at it.

Mrs. LUMMIS. Thank you. Anyone else care to comment on that?

Mr. UPTON. My understanding is that the increased flows have not helped the smelt population, and there are a lot of factors involved with it. George just mentioned a few, but also I don't think we really understand completely the science of where habitat exists for smelt, or the predators or any of that. So we really don't know. And the fact is that throwing more water at it hasn't helped, in my understanding.

Mrs. LUMMIS. So should there be a concept in the Endangered Species Act of a futile call on water? In Wyoming, we have this concept of a futile call. If delivering water to a senior water rights holder downstream will not even get there because the conveyance system would cause the water to just seep into the ground, so the water can't be delivered, then you can't make a call on a senior right because it is futile, it won't get there.

Should there be a similar concept in the Endangered Species Act where if it is documented, as this is, that providing more water, cutting off other water uses to provide more water does not, in fact, aid the recovery of a species, then there is no sense doing it? Does that make sense?

[Applause.]

Mrs. LUMMIS. Yes, Mr. Coleman?

Mr. COLEMAN. One of the problems is they never set a standard or criteria for what a successful delta smelt population is, so they don't know.

Mrs. LUMMIS. That is a problem that Chairman Hastings and I identified through over a year of hearings in Washington regarding some of the flaws of the Endangered Species Act, is that if you can't measure successful recovery, how do you ever get a species delisted? So, thank you for that. That is another confirming bit of testimony that we are on the right track with regard to requiring scientific data and transparent data as criteria for recovery of a species so we can improve the success rate.

Right now, there is only a 2 percent or less success rate in recovery of species listed under the Endangered Species Act. Well, that is a failed law.

[Applause.]

Mr. COLEMAN. The smelt population this season is the second lowest it has been, so it has been a total failure. But don't lose sight. It is not about the smelt. That is just a lever to take our water. But it has not been successful.

[Applause.]

Mr. DELGADO. May I add another thing?

Mrs. LUMMIS. Yes, sir, Mr. Delgado. By the way, I loved hearing your background. It mirrored my own. So thank you for——

Mr. DELGADO. I thank you very much. I am not the only one that has that background. There are a lot of west side farmers that come from different parts of the country—Oklahoma, Kansas. My history goes back to the valley for all of my life. I have always respected and will always have respect for farmers in the west side and any farmer that farms anywhere in the world because it is not as easy as it looks. We make it look pretty sometimes in the fields, but when you are really out there trying to make decisions, it is tough enough making decisions when we are dealing with markets and labor and dealing with all these environmental laws that have been imposed on us. So I am very thankful that I have had the opportunity to be part of the west side. But unfortunately, it is very sad because the west side is starting to die already.

The CHAIRMAN. The time of the gentle lady has expired.

Mrs. LUMMIS. Thank you, Mr. Chairman. I yield back.

The CHAIRMAN. I recognize Mr. Nunes.

Mr. NUNES. Thank you, Mr. Chairman. Mr. Chairman, thank you for yielding.

Ms. Beland, just for the record, the Governor opposes the House-passed legislation?

Ms. BELAND. Yes.

Mr. NUNES. So what is the Governor's plan? Does the Governor believe that there has to be any Federal laws that are changed or modified in order to bring water to all Californians?

Ms. BELAND. The Governor believes we need to work on the infrastructure projects that he has put forward in both the Water Action Plan and the Bay Delta Conservation.

Mr. NUNES. You could build tunnels, you could build canals, you could build dams, but the problem remains that if the Federal laws aren't changed, those could all go unused like they are today.

[Applause.]

Mr. NUNES. I am just trying to see, do you agree with that statement I just made? Are we using the infrastructure that we have already built in the last 100 years? Are we using it to its fullest today?

Ms. BELAND. I don't think so.

Mr. NUNES. OK. That is the right answer.

[Laughter.]

[Applause.]

Mr. NUNES. So what dams does the Governor support at this time to construct? Can you name any dams that he would actually support raising, or new ones?

Ms. BELAND. Well, we are looking at a partnership right now with Sites Reservoir up in the Northern Sacramento Valley that will provide us with an opportunity to build in conjunction——

Mr. NUNES. But just one, just the one.

Ms. BELAND. That is one example.

Mr. NUNES. Ms. Beland, you are a young lady, so this comment is not toward you, but the Governor was Governor of this State 40 years ago when I was born. If the Governor hasn't figured out what new storage projects have to be built, then I think that any storage that is talked about—and I think everybody in this crowd needs to understand this—that it is only used to pacify those of us who have been fighting for storage for a long time.

[Applause.]

Mr. NUNES. Would the Governor be willing to waive SEQRA to construct dams in this State?

Ms. BELAND. I can't speak to that. I don't know.

Mr. NUNES. OK. But the Governor does waive SEQRA, sign bills to waive SEQRA with the basketball stadium arenas, football stadiums? You know about that, right?

Ms. BELAND. Yes.

Mr. NUNES. Mr. Murillo and Ms. Beland, could you comment on—so the graph I have on the screen, it is basically 2.5 million acre feet that we are short, on average, in this region, basically from Merced to Bakersfield. Would you agree with those numbers?

Mr. MURILLO. I don't know if I can agree with those exact numbers, but I believe we are short.

Mr. NUNES. Is it close to 2.5 million acre feet, or is it——

Mr. MURILLO. I don't know. I can't respond to that. What I can tell you is we do understand that we are short. We are trying to deliver water to Ag, and we do understand that we have a number of regulatory requirements that we are trying to meet.

This year, what we have been doing is we have been working with the district general managers and we have been coming up with ideas this year. Last year, what we dealt with is basically trying to operate at minus-2,500, minus-2,000. This year we have moved forward and we are trying to operate at minus-5,000, minus-6,600. So we are trying to push those Biological Opinions to the limit.

Mr. NUNES. No, I understand, Mr. Murillo. I mean, you are in charge of running the Bureau. I am not asking you to be a water expert, but we have to identify the problem. That is one of the challenges that I have realized in my time fighting this issue, is that we always talk in the abstract because it is easy for politicians to stand up here and say, oh, I want to build 20 dams, I want to build every dam imaginable. They always say that when they campaign, but then no dams ever get built.

So I am trying just to get you and your folks and the Governor's folks and us, the Congress, to focus on what is the shortfall. I mean, it has to be close to 2.5 million acre feet, right?

Mr. MURILLO. Like I said, I don't know what the exact number is. I know we are short.

Mr. NUNES. I mean——

Mr. MURILLO. Well, you are asking me the question.

Mr. NUNES. We are not this much short, right?

Mr. MURILLO. Exactly.

Mr. NUNES. We are a lot more than that.

Mr. MURILLO. Yes, we are more than that. I agree with that.

[Laughter.]

[Applause.]

Mr. NUNES. I think it is really important, and I am asking both of you, please, to submit after this how much the region is short. You can take this graph with you. If you come back with 2 million acre feet, something like that, we can discuss it. But we have to figure out what the real problem is in order to build a solution.

The CHAIRMAN. Would the gentleman yield real quickly?

Mr. NUNES. Absolutely, Mr. Chairman.

The CHAIRMAN. How quickly can you get that information to Mr. Nunes? I just ask. How quickly can you get that for the committee?

Mr. MURILLO. We should be able to get it within a couple of weeks.

The CHAIRMAN. A couple of weeks?

Mr. MURILLO. Yes.

The CHAIRMAN. All right, we will give you that leeway, 2 weeks from today, which would be about April 2, I think, if my memory serves me correctly. So the committee can expect that information on April 2. Do you both concur?

Ms. BELAND. Yes.

The CHAIRMAN. Affirmative on both cases.

I yield back.

Mr. NUNES. And we will get in writing the number we are asking for.

The CHAIRMAN. The time of the gentleman has expired.

Mr. NUNES. Thank you, Mr. Chairman.

The CHAIRMAN. I recognize the distinguished Majority Whip, Mr. McCarthy.

Mr. MCCARTHY. Thank you very much, and thank you to all the witnesses.

A couple of things I would like to focus on, one following up on what Congressman Nunes said. We all agree in storage, but if we can't move the water through the delta, what will the storage matter?

It goes back to the key question that a lot of people here said. When you answered the question that the Governor opposed a bill that passed the House, my question is what part of it does he oppose? Do you know? I will come back to you.

I want to put the personal aspects of this because we do have some cameras here and we have some people from out-of-state as well as part of this conference bill. What does that really mean that is affecting us today? I listened to Mr. Starrh say 1,000 acres. This isn't a rotation crop. You said you grow almonds.

Mr. STARRH. Right. We are taking out—we are going to stop irrigating 1,000 acres of trees that are anywhere from 15 to 18 years old. So they are productive trees, and with that we won't be able to—you don't harvest them, you don't hire the people to harvest them, you don't do any of that. None of that happens. The product is gone. The trees die. And you have invested 14 years of capital to get them built. You put them in, it takes 3 years before you get even a little bit of a crop off an almond tree, and then, yes, it is gone.

Mr. MCCARTHY. These trees are 14 years old. You are fallowing them. They are dying.

Mr. STARRH. Right.

Mr. MCCARTHY. You can't bring them back the next year.

Mr. STARRH. Right, right.

Mr. MCCARTHY. A thousand acres.

Mr. STARRH. A thousand acres.

Mr. MCCARTHY. I want to put another face, not to the farm but to Mayor Chavez. You talked, even the last time, you gave some numbers in your community, a little less than 7,000, over 90 percent Hispanic, working, many working in the fields and others, and you said that you had 40 percent unemployment the last time. Can you tell us some of the aspects of what happened during that 40 percent unemployment?

Mayor CHAVEZ. During that time we had long lines for food. Our crime rate went up during that time that there was no water. They had cut down on a lot of jobs. A lot of our community members had moved away, too, because of trying to find work somewhere else.

Coming from the community, from Huron, I have been there all my life. I worked in the fields. I picked cotton. I picked grapes. I know how hard it is for these people to try to have to look for work, and they have to travel. Some of them have come back, but some have moved away and stayed away, and that affects our community a lot, and the revenues themselves and everything else that we have to pay.

A lot of our homes went empty. A lot had to move out of their homes, and those went into foreclosure. So we lost that revenue, too.

Mr. MCCARTHY. So, the entire city affected, from the police, fire, school districts as well, because of the unemployment?

Mayor CHAVEZ. Exactly. Because some couldn't pay for their homes, couldn't pay the property tax, we lost that money. So the city was hit very hard with that in 2000—that was in 2009. We are looking for more shortage this year.

Mr. MCCARTHY. So those are some of the aspects that happen rather quickly that we are in the mode of happening right now.

If I could turn back, and I don't hold it to you. You are not directly with the Governor. I know Secretary Laird couldn't be here. I served with him in the Assembly. I was hoping he could have been. I know the Governor declared this an emergency, so I was hopeful that the Secretary could be here.

Part of why this hearing took place, I know when the President came out—and correct me if I am wrong, Mr. Chairman—the entire Democrat delegation inside Congress requested a hearing, maybe not from this committee but requested a hearing. And did you invite everyone?

The CHAIRMAN. If the gentleman would yield, I sent a letter to everybody. Once we established the hearing date, I sent a letter to every Democrat Member of Congress in both the House and the Senate asking them to come to this hearing.

Mr. MCCARTHY. If I could turn back to the Governor's difference of opinion on a bill that has passed the House dealing with the water issue, and none through the Senate, does he have some specific reasons that he would oppose it?

Ms. BELAND. I think two main concerns: one, that we are putting one set of interests above another.

Mr. MCCARTHY. And which would that be? Would that be humans above fish?

[Applause.]

Mr. MCCARTHY. What else?

Ms. BELAND. And it would interfere with the flexibility that we would need to respond to the current crisis. Those were the two reasons that he gave in his letter.

Mr. MCCARTHY. If I could turn to Mr. Upton, I thought you gave very interesting testimony, being part of that negotiation and restoration and coming back with the environmental accountability, you said. And the one thing I heard from all this, many times we have gone into these agreements, and what people have some agreement for is not what turns out to be. I find, yes, I want more storage, but if we can't get that water through the delta, we are not utilizing the capacities we have now.

Maybe you can elaborate a little on what you were meaning by that based upon the negotiation and the restoration of others of where it ended up, and the accountability that you think you need in the environmental as well.

The CHAIRMAN. Real quickly, Mr. Upton, real quickly.

Mr. UPTON. OK. All right. Well, we had an agreement, a signed agreement that they were supposed to help us get our water back, and they have reneged on that, so we are not going to get our water back. You are talking 250,000 acre feet, approximately, every year, on average, that is going to be sent down the San Joaquin River, and that means we are going to have to pump from underground to make it up. That is unsustainable. That is unsustainable. So what is going to happen? You are going to have to idle acres. That is the only other solution, unless you are going to rewrite this law for a warm-water fishery and have the Temperance Flat.

This situation could be addressed, but it can't be as long as they continue to listen just to the environmental side of this.

[Applause.]

The CHAIRMAN. The time of the gentleman has expired.
I recognize Mr. Denham.
Mr. DENHAM. Thank you, Mr. Chairman.
Mr. Knell, in your testimony you talked about past years being able to transfer water to Westlands Water District. In 2012, you were prohibited from doing that. Can you explain why?
Mr. KNELL. The timing of the water, when it was available, and the capacity at the pumps being consumed for other purposes just delayed the pumping long enough that the water couldn't be released. We are trying to release the water within a timeframe that was fish-friendly so we could benefit the fisheries in our river, in the Lower San Joaquin. But there again, there just wasn't any capacity at the pumps due to pumping of other water. That is why we needed the storage at New Melones. Had we been able to back that water up the hill and wait until that capacity was available, then we could shoot the water across.
Mr. DENHAM. So you were prohibited in 2012. What happened in 2013? Were you able to pump in 2013, the west side farmers?
Mr. KNELL. No. Our district has no surplus water available for transfers due to the depth of this drought at this time.
Mr. DENHAM. And this year as well, I assume, there will be no transfers——
Mr. KNELL. I am sorry, I am sorry. In 2013, we moved 80,000 acre feet. I am sorry. I apologize. We moved 80,000 acre feet across the delta last year. This year I was saying we have no water for movement.
Mr. DENHAM. Mr. Murillo, when Oakdale Irrigation District has water to transfer, is there anything prohibiting them from addressing that?
Mr. MURILLO. I think, as you mentioned, we look at a number of factors. We look at the demands that are required.
Mr. DENHAM. Are you currently addressing this problem?
Mr. MURILLO. We are taking a look at it. But like I said, we look at a number of factors. We look at the senior water rights that are in there. We also——
Mr. DENHAM. When you say you are taking a look at that, what does that mean to the farmers in my community? Are you looking at it so that if there is excess water they can plan on having that, or is this something that is going to take several years to look at and do a study?
Mr. MURILLO. I think, as we mentioned before, last year we moved water. What we do is we take a look at what the conditions are. We take a look at what their proposals are. And we take a look to see what the impacts to the CVP overall are. So we consider that, and if we think that the conditions are right, we can go ahead and move that water through.
Mr. DENHAM. And you work with Oakdale Irrigation District on that?
Mr. MURILLO. Yes, we work with all districts.
Mr. DENHAM. Mr. Knell, I introduced H.R. 2554, which would allow more storage at New Melones Reservoir, again at no cost. I have heard about Reclamation seeking all avenues to expand and improve water supplies due to this drought. Has Reclamation contacted you about this bill?

Mr. KNELL. I am sorry?

Mr. DENHAM. Has Reclamation, have they contacted you about the bill for New Melones?

Mr. KNELL. No. To my knowledge, no.

Mr. DENHAM. Have you heard of the bill?

Mr. MURILLO. I have not seen the bill myself.

Mr. DENHAM. OK. We will get you a copy of that.

Mr. MURILLO. Thank you.

Mr. DENHAM. It is certainly alarming that you wouldn't be focusing on all water bills, certainly in an emergency, where the State has declared an emergency. The President has now declared an emergency, and we are not looking at all of the bills that would deal with water storage, especially one that would come at no cost.

So will you commit to this panel today that you will not only take a look at the bill but you will actually sit down with Oakdale Irrigation District and work on a no-cost solution with new water storage in New Melones?

Mr. MURILLO. I personally will take a look at the bill, and we can sit down and have a discussion.

Mr. VALADAO. If the gentleman would yield just for a quick second?

Mr. DENHAM. I will yield.

Mr. VALADAO. I just want to make sure for the record, Mr. Denham, that indeed your bill passed the Congress, and I am not sure about Mr. Murillo. I am not sure that they commented on the bill, but I know that the Governor's Office opposes the bill, because they have sent a letter opposing the legislation.

Mr. DENHAM. I understand that the Governor, Ms. Beland, opposed this bill. Does he oppose the water storage at no cost at New Melones Reservoir?

Ms. BELAND. I don't have the background on the Governor's opposition to your bill in particular. I can respond back to you.

Mr. DENHAM. I would look forward to a response.

As well, there are a number of different water storage projects, both in authorization as well as new storage in the bill.

Look, here is part of my frustration.

The CHAIRMAN. Would the gentleman yield? Would the gentleman yield?

Mr. DENHAM. You have to show us what you are for. If you are having an emergency here and you disagree with an approach, then either introduce a bill of your own or show us what you are for. But we can't continue to negotiate with ourselves.

[Applause.]

Mr. DENHAM. I would be happy to yield to the Chairman.

The CHAIRMAN. I would ask the Undersecretary, when will you respond to him, to his specific question? When can he expect that response? It is an important response. We need to find out what the State of California wants. He says it very well. When will they respond?

Ms. BELAND. I can get back to you by Friday.

The CHAIRMAN. When will you respond?

Ms. BELAND. I can respond by Friday.

The CHAIRMAN. By Friday?

Ms. BELAND. Yes.

The CHAIRMAN. You will get a response to Mr. Denham's question?

Ms. BELAND. Yes.

The CHAIRMAN. All right. I yield back to the gentleman. He still has some time because I took it.

Mr. DENHAM. Thank you. I had a number of questions for the State Water Resources Control Board, Mr. Chairman. I will submit those questions through the committee. I would expect answers.

But I, as Mr. Upton, I find it alarming as well that the NRDC continues to sue on any new water storage, but yet the Governor, who is now claiming that we have an emergency here, continues to appoint the very same people that are suing us on all of our water projects.

[Applause.]

The CHAIRMAN. The time of the gentleman has expired.

I recognize Mr. Valadao for 5 minutes.

Mr. VALADAO. Thank you, Mr. Chairman.

My question, my first question, goes to Ms. Chavez in Huron. The last year, what was your water allocation for the city?

Mayor CHAVEZ. The last year the water allocation was 1,125. This year we received a letter from the Bureau of Reclamation that we will receive 649 acre feet. That makes us short 476 acre feet, and that is what we have to deal with this 2014–2015 year.

Mr. VALADAO. One of the most frustrating things for me, because I represent the area in Congress and I have had a lot of time with the mayor in her district office there, is you have papers coming from the Bureau saying they get 70 percent of historical use. They like to throw those big numbers out there, you are getting 70 percent, but they put the "historical use" in there in small print, and it ends up being where you get less than 50 percent of your actual water, and you are putting these constituents in real danger. It is not just about the farmers but it is also about people in these communities. They are very important to us. They are important to everybody in this area.

My next question, Mr. Delgado, you started this farm. You are a new farmer. How do situations like today—I mean, you don't have generations behind you of owning farmland. You started this on your own. You started as a farm worker. You built your way up. You probably don't have the same capital as someone who has been around for two generations or three generations.

How does this affect you when you have to compete with these other farmers, and how does that affect you in achieving the American Dream, what we are all here for?

Mr. DELGADO. Well, personally, I have been very blessed to have this opportunity. So far, I have been blessed with—I have very close ties with business people, farmers who have been around for three or four generations. They basically trust me with my word, that my word is good, that I am honest, I have character, and that I have good intentions with the land.

I have also been blessed that I have a family that supports me. My wife has a doctorate in education. I have a son who is a water law attorney. I have another son who teaches math. I have my daughter, who is a physical therapist. All of their wives are college educated. So I have been blessed with what is more important, which is really a family that comes first.

If something were to go wrong, if I can't survive, well, I will be a statistic like anybody else. I am gone. But I am going to try to hang on like everybody else is.

Mr. VALADAO. One more question. Mr. Starrh, when you rip out those 1,000 acres of trees, just like a lot of other farmers are doing, does it require—obviously, you have an investment in those trees. You probably have loans. You probably have other acreage that you can borrow against to help carry yourself through those tough years. How long does it take you to rebuild those 1,000 acres, and how long does it affect other farmers in the area the same?

I mean, obviously, 14 years you have had some of those trees, 18 years, but it takes now many years to actually get to the point where you rip those trees out that there was investment in, hopefully your loans are paid off, before you can plant the next group of trees and get those into production where you can start actually making payments to the bank for it.

Mr. STARRH. Right. Well, when you take them out, then you would have to lay it out for a couple of years, and then you start the process. You plant, and then you have another 3 years. But, David, the key is if you don't have water coming in the future, the decision will be there will be no trees and there will be no work, and there will be nothing. I mean, we will just go back to—I mean, I will find a place in Texas, I guess.

[Laughter.]

Mr. VALADAO. So that goes to Mr. Delgado. When someone like my family and Mr. Delgado's family starts a business and they have to borrow and they have to do everything they can—you still work on another farm, I understand?

Mr. DELGADO. Right now, anyone—well, I can't speak for everyone. I speak for myself. We made major investments before this 1992 Improvement Act was put in place. So we purchased—we were able to buy land. We bought land, and it has to be financed with banks, and the banks want answers too. The banks, when they start having uncertainty, they start pulling back, and they have that fear. Myself, I have invested money into land. I purchased land, and at that time it did have 100 percent water allocation. I thought we were in pretty good shape with water. And yet I understood that the drought would affect it, like most farmers know that.

Now, another issue is we have invested also in trees, like Larry has, and drip irrigation systems. Every acre that I know on the west side and people that use Federal water, basically it is all under drip irrigation. There is not any water wasted.

Mr. VALADAO. So you have this investment. It is all at risk. We have communities that are affected by this. We are looking at opportunities, or people are looking for other opportunities to move. Communities are out of water, and I heard in your comments earlier, Ms. Beland, that you are trying to keep the salinity levels at a controllable level in the delta, and I understand the concerns there.

But throughout the year, there were a lot of timed releases from these reservoirs that are paid for by the farmers, a lot of farmers here in the valley. When these reservoirs are releasing water to help the salinity levels in the delta, it is interesting that our money, our infrastructure that is supposed to be protecting jobs, supposed to be protecting communities' clean water, is concerned with something that, if those dams weren't there, the water would have just gone out to the ocean, the salinity levels would come in anyway. It is nature. It happens. It is not something that you should be using this type of infrastructure for.

We are not God here. We are not concerned—we can't prevent water from going into these communities and hurting these people in the infrastructure that is there, all in the name of salt or in the name of a fish, because it just doesn't make any sense.

The salinity levels, if you truly wanted to go back to nature, tear down the dams, let the water go out when it melts and then it is over, and by the end of summer the salt comes in anyway.

[Applause.]

Mr. VALADAO. You can't use infrastructure like this to say that you are trying to protect the environment where, if everything was in its natural state, then the environment would be at risk, and then it would recover, just like it probably happened before humans came to California and started doing all these things.

It is just interesting for me because when you were making those comments earlier, you are going down this direction where you are trying to use things that aren't natural to protect something that is natural, and it is the most asinine thing I have ever seen in my entire life.

[Laughter.]

[Applause.]

Mr. VALADAO. And it has been good because we obviously have a great panel here.

I wanted to end because I know that I am the last one, and I am rushed because of the time limit. But I wanted to say thank you to the city of Fresno today for allowing us to do this. This was a really good event.

[Applause.]

Mr. VALADAO. We had a lot of staff from Fresno City Hall help us a lot today, and I want to thank the Fresno City Hall staff for being so helpful with our staff.

I also want to thank the Fresno PD. I hear they were really nervous about the size of the crowd, and I am glad we had a crowd to make them nervous.

[Laughter.]

Mr. VALADAO. But Selma High School ROTC, thank you for taking the time. I know they have already left. They took their flag with them. We were able to give them a flag flown over the capital.

[Applause.]

Mr. VALADAO. They are in the back. Thank you.

Reverend Baptista, thank you for taking some time for us today.

William Bordeaux, thank you for doing the Pledge.

And Chairman Hastings, for convening this hearing today, it means a lot to us here in the valley.

Congresswoman Lummis, thank you very much for making the trip from Wyoming. Everybody here really appreciates your comments today, and your questions.

[Applause.]

Mr. VALADAO. And just on a closing line for myself, I have been in Congress for a year now, and this fight has been going on much longer. I was dragged into politics really the last time they had a huge rally on the west side that Devin was a part of in the middle of the orchard. I don't know if you remember that, in September of 2009. I never considered myself somebody to run for office.

Having situations like this and being able to be in the fight on this side of the dais and having a team like this that is going to work together to pass legislation—when you pass legislation, what it means is you are putting your money where your mouth is. You are putting forth your ideas, what you want to do, how you are going to fix this problem.

But you need partners. For us to do what we did so far has been really great. But the next step has to go through for us to actually deliver.

So we have to continue this fight. We have to continue to move forward, and this is not over just because this is over today. We are going to continue to fight in Washington, and thank you so much for being a part of this.

[Applause.]

The CHAIRMAN. I thank the gentleman.

And I want to thank all of the witnesses for your testimony here. From time to time there are questions on follow-up. If you get asked by a Member, please, I hope you will respond in quick time.

I will say this. When I ask for timelines, I get very frustrated as a committee chairman when we ask people to respond and they don't respond. So I expect the State and the Bureau to respond in the time that we agreed upon here on the public record.

With that, I thank everybody for coming.

There being no further business, the committee stands adjourned.

[Applause.]

[Whereupon, at 12:38 p.m., the committee was adjourned.]

[Additional Material Submitted for the Record]

BUREAU OF RECLAMATION, MID-PACIFIC REGION, APRIL 2, 2014

RESPONSE TO HOUSE COMMITTEE ON NATURAL RESOURCES FIELD HEARING QUESTION ON SOUTH-OF-DELTA WATER SUPPLY DEFICIT

INTRODUCTION

At the March 19, 2014 House Committee on Natural Resources field hearing in Fresno, California, a slide was presented by the committee showing a 2.4 million acre-foot "deficit" to water supplies south of the Sacramento-San Joaquin Delta in California (this is referred to as the "Deficit Slide" elsewhere in this document). The total deficit was calculated as the sum of "deficits" created by requirements and water supply demands classified in several different categories, including Central Valley Project Improvement Act [CVPIA] environmental water, CVPIA refuge supplies, the Biological Opinions [BiOps] for the Central Valley Project [CVP] and State Water Project [SWP], groundwater overdraft, and the San Joaquin River Restoration Program [SJRRP]. Reclamation and the State of California were requested to provide their own representation of the water supply challenges and shortages to users south of the delta. This document responds to that request. In developing this response, Reclamation relied upon data from the State of California's Department of Water Resources.

CENTRAL VALLEY PROJECT DATA

For reasons discussed later in this document, Reclamation does not agree that the figures shown in the Deficit Slide form an accurate characterization of a water supply ''deficit'' to water demands south of the delta. As further detailed in this document, there are many factors which make it difficult to develop a quantitative analysis of water supply challenges and shortages to water users south of the delta. Given these difficulties, Reclamation believes the best way to quantitatively illustrate the water supply challenges south of the delta is through a display of the recent history of allocations and supplies to CVP contractors both north and south of the delta (Table 1).

Of particular interest to the committee in reviewing Table 1 will be the allocations to south-of-delta agricultural contractors and the Friant Division. These allocations illustrate the amount of the total contract volume to those water users that Reclamation has not been able to serve over the past two decades due to a large number of hydrologic and regulatory controlling factors which vary (sometimes dramatically) from year-to-year.

When reviewing the attached table, it is important to note several other key issues and factors. These key issues and factors highlight the difficulties previously noted in creating a single ''deficit'' or water supply shortage value. The issues and factors include:

- Hydrologic conditions are a significant factor in annual allocations and south-of-delta supplies. Table 2 illustrates the precipitation indices for the Sacramento and San Joaquin basins, as well as the Sacramento River Basin Year type (calculation methodologies for these indices were developed by the State Water Resources Control Board to summarize hydrologic conditions as part of their regulatory activities). The precipitation indices are shown in relation to the final south-of-delta agricultural allocation for each year.
- Previous-year hydrologic conditions also play a strong role in annual water supplies to the Central Valley. To illustrate this, Table 2 also shows the carry-over storage totals in CVP reservoirs.
- The water service allocations shown in Table 1 do not reflect the complete annual water supply to many water users south of the delta. Many water users secure and develop other water supplies from local sources (groundwater, local runoff, reuse) and through other means such as transfers, agreements, exchanges, and rescheduling of previous-year CVP water supplies.
- Specific to the Friant Division of the CVP and as noted in the footnotes on Table 1, the Friant Class 2 allocation is considered a ''supplemental'' water supply to the Friant Division, and should not be considered a firm or ''base'' water demand for the purposes of analyzing a water supply shortage or deficit to south-of-delta supplies.
- Attempting to create a single summary statistic or ''deficit'' by averaging the allocation and water supply values in Table 1 would be problematic due to the dynamic nature of recent changes in regulatory requirements, particularly when coupled with highly variable hydrologic conditions.

STATE OF CALIFORNIA DATA

In addition to Table 1 which displays the effects of water supply challenges on water service to CVP contractors, Reclamation would like to direct the committee to products developed by the State of California which summarize the reliability of deliveries to the SWP. These reports can be found on the State's Web site at: *http://baydeltaoffice.water.ca.gov/swpreliability/.*

The most recent SWP reliability report was published in June 2012. The document is currently being updated and is presently available in ''public review draft'' on the State's Web site at: *https://msb.water.ca.gov/documents/86800/202762/DRR2013_Report_20131210.pdf.*

Pertinent to the committee's request, the analysis of ''recent trends in SWP delta exports and Table A deliveries'' found in Chapter 4 of the public review draft illustrates estimated reductions in SWP exports and deliveries as a result of changes over time to the operating environment of the SWP. These changes are shown graphically in Figure 4–1 of the public review draft (copy of Figure 4–1 attached).

In addition to the SWP Delivery Reliability Reports, the committee may be interested in reports produced by the State summarizing water supply and uses across the State. The reports are produced every 5 years in conjunction with updates to the California Water Plan (Bulletin 160), as required by the California Water Code. The 2013 Update of the California Water Plan is presently available in ''public review draft'' form on the State's Web site at: *http://www.waterplan.water.ca.gov/cwpu2013/index.cfm.* The final updated Plan will be published by summer 2014.

The California Water Plan includes regional reports which provide details on the water supplies, water uses, and overall water balance in each region of the State. Pertinent to the Committee's request are the San Joaquin River and Tulare Lake regional reports. The most recent final reports on these areas (2009 Update) are available on the State's Web site at these links: *http://www.waterplan.water.ca.gov/docs/cwpu2009/0310final/v3_sanjoaquinriver_cwp2009.pdf* and *http://www.waterplan.water.ca.gov/docs/cwpu2009/0310final/v3_tularelake_cwp2009.pdf.*

Of particular interest may be Figures SJ–3 and TL–21 within the two regional reports, which are water balance charts which illustrate the wide variety of supplies to the area beyond those provided by the CVP and SWP. Also attached to this document are Figures SJR–20 and TL–16, the 2013 updated versions of these charts, which are available now in public review draft form. These charts highlight one of the key complications with developing an analysis of the water supply shortages south of the delta, which is that groundwater extraction is often used in the place of Federal, State, and local surface water supplies in years when surface supplies are reduced. This can result in increased extraction of groundwater in the basin.

If the increased groundwater extraction continues and is not naturally or artificially replenished over time, it may lead to overdraft. Groundwater overdraft is defined as the condition of a groundwater basin in which the amount of water withdrawn by pumping exceeds the amount of water that recharges the basin over a period of years, during which the water supply conditions approximate average conditions. In contrast, declining storage over a relatively short period of average hydrologic and land use conditions does not necessarily mean that the basin is being managed unsustainably or is subject to overdraft. Utilization of groundwater in storage during years of diminishing surface water supply, followed by active recharge of the aquifer when surface water or other alternative supplies become available, is a recognized and acceptable approach to conjunctive water management.

Developing an accurate depiction of groundwater overdraft, and particularly how it relates to shortage of surface water supplies, is a very complex task. For this reason, for the purposes of the present request, the committee may be best served by utilizing the data found in Figures SJR–20 and TL–16 to observe the variations in water supplies and use for the time period 2001 to 2010 in the San Joaquin River and Tulare Lake regions, and the aforementioned SWP Delivery Reliability Report to estimate the challenges in serving south-of-delta water supplies to SWP contractors.

ISSUES WITH ''DEFICIT SLIDE''

As previously noted, Reclamation believes the information provided above and in the attached tables and charts are currently the best way to quantitatively represent water supply challenges to south-of-delta water users. For the benefit of the committee, Reclamation would like to provide the following additional notes as to why it believes the Deficit Slide presented at the hearing does not accurately capture or analyze these challenges.

In general, the total ''deficit'' shown on the slide at the hearing appears to incorrectly add together several different categories of water supply requirements and demands under the CVPIA, Endangered Species Act [ESA], and SJRRP in order to calculate a total amount of water ''lost'' to south-of-delta water users. Reclamation questions the appropriateness of the use of several of the values for ''deficit'' shown in the slide, as well as their magnitude, as follows:

- The numbers cited cannot be considered additive, because portions of the water used for the requirements listed on the slide are also utilized for other purposes, including supply to south-of-delta water users. For instance, the amount listed on the slide for CVPIA environmental water (800,000 acre-feet) includes:
  - Releases from upstream storage that, at times, can be exported for use south-of-delta;
  - Foregone pumping that meets the requirements of the BiOps or State water quality requirements (which are separately listed as a ''deficit'' on the slide).
- 250,000 acre-feet of ''deficit'' is ascribed to the SJRRP. The parties to the Settlement in *NRDC* v. *Rodgers* spent numerous years negotiating procedures for determining and accounting for the water supply impacts of the SJRRP. The procedures are documented in the Restoration Flow Guidelines, dated December 2013.

- Using the procedures agreed to by the parties, the average long-term water supply impact to the Friant Division long-term contractors as a result of the Settlement is estimated at 185,000 acre-feet per year.
- This amount does not account for the recirculation, replacement and offset actions that Reclamation is implementing as part of the Settlement. With implementation of the recirculation, replacement and offset actions, the actual average long-term water supply impact will be less. However, these actions will vary from year to year and are not possible to calculate at this time.

- The amount listed on the slide for CVPIA refuge supplies (400,000 acre-feet) is larger than the amount delivered to south-of-delta refuges in recent years (generally approximately 250,000–270,000 acre-feet), and a portion of the water delivered to refuges generally returns to rivers where it can be used again for other purposes (pertains to refuge deliveries both north and south of the delta).
- Reclamation is not certain as to the source or accuracy of the data of some of the other values illustrated on the slide, such as the amount of water ''loss'' for the BiOps, or the ''groundwater overdraft'' value.
  - It is also not clear to which water users the ''groundwater overdraft'' value applies; any groundwater overdraft quantified for the area may apply to more than just CVP water users, and as such, attempts to resolve any such issue may involve more than operations of the CVP.

SUMMARY

It is the belief of Reclamation that the information provided through the attached tables and the State's reports serve to respond to the Committee's request for information on the quantification of water supply challenges to users south of the delta. Should the Committee need additional information or wish to discuss these numbers, please contact Reclamation's Congressional Affairs representatives.

Table 1

Central Valley Project Water Supplies, Allocations, and Refuge Deliveries; 1992-2013

	Senior Water Rights				North of Delta				South of Delta				Friant Division				Eastside[5]		Refuge Deliveries[9,10,11]			
Year	SRSC[2]	Volume[1] (AF)	Exchg. Contrctrs.	Volume[1] (AF)	Agric. Alloctn.	Volume[1] (AF)	M&I[3] Alloctn.	Volume[1,4] (AF)	Agric. Alloctn.	Volume[1] (AF)	M&I Alloctn.	Volume[1,5] (AF)	Class 1 Alloctn.	Volume[1] (AF)	Class 2 Alloctn.[6]	Volume[1] (AF)	Alloctn.	Volume[1] (AF)	Level 2 Delivered	Volume (AF)	Inc. Level 4 Delivered	Volume (AF)
1992	75%	255,083	75%	630,000	25%	111,851	75%	151,984	25%	487,802	75%	121,268	83%	664,000	0%	0						
1993	100%	340,111	100%	840,000	100%	447,402	100%	335,338	50%	975,604	75%	121,268	100%	800,000	90%	1,261,328			71%	300,010	8%	10,550
1994	75%	255,083	75%	630,000	35%	156,591	75%	151,984	35%	682,923	75%	121,268	80%	640,000	0%	0			71%	300,700	7%	9,523
1995	100%	340,111	100%	840,000	100%	447,402	95%	318,571	100%	1,951,208	100%	161,690	100%	800,000	100%	1,401,475			78%	328,553	30%	39,355
1996	100%	340,111	100%	840,000	100%	447,402	100%	335,338	95%	1,853,648	100%	161,690	100%	800,000	58%	812,856			88%	371,127	20%	26,106
1997	100%	340,111	100%	840,000	100%	447,402	100%	335,338	90%	1,756,087	90%	145,521	100%	800,000	60%	840,885			90%	379,135	28%	37,780
1998	100%	340,111	100%	840,000	100%	447,402	100%	335,338	100%	1,951,208	100%	161,690	100%	800,000	10%	140,148			85%	359,874	34%	44,967
1999	100%	340,111	100%	840,000	100%	447,402	95%	318,571	70%	1,365,846	95%	153,606	100%	800,000	20%	280,295			82%	345,221	34%	45,121
2000	100%	340,111	100%	840,000	100%	447,402	100%	335,338	65%	1,268,285	90%	145,521	100%	800,000	17%	238,251	58%	89,900	84%	353,800	50%	66,105
2001	100%	340,111	100%	840,000	60%	268,441	85%	172,248	49%	956,092	77%	124,501	100%	800,000	5%	70,074	22%	34,100	87%	365,974	48%	63,322
2002	100%	340,111	100%	840,000	100%	447,402	100%	335,338	70%	1,365,846	95%	153,606	100%	800,000	8%	112,118	8%	12,400	91%	383,842	61%	80,970
2003	100%	340,111	100%	840,000	100%	447,402	100%	335,338	75%	1,463,406	100%	161,690	100%	800,000	5%	70,074	6%	9,300	93%	391,635	58%	77,053
2004	100%	340,111	100%	840,000	100%	447,402	100%	335,338	70%	1,365,846	95%	153,606	100%	800,000	8%	112,118	0%	0	91%	385,731	51%	68,262
2005	100%	340,111	100%	840,000	100%	447,402	100%	335,338	85%	1,658,527	100%	161,690	100%	800,000	UC[7]		28%	43,400	92%	388,803	62%	83,211
2006	100%	340,111	100%	840,000	100%	447,402	100%	335,338	100%	1,951,208	100%	161,690	100%	800,000	UC[7]		100%	155,000	90%	380,072	68%	90,545
2007	100%	340,111	100%	840,000	100%	447,402	100%	335,338	50%	975,604	75%	121,268	65%	520,000	0%	0	29%	44,950	92%	388,521	34%	45,049
2008	100%	340,111	100%	840,000	40%	178,961	75%	151,984	40%	780,483	75%	121,268	100%	800,000	5%	70,074	23%	35,650	91%	386,181	27%	35,566
2009	100%	340,111	100%	840,000	40%	178,961	75%	151,984	10%	195,121	60%	97,014	100%	800,000	18%	252,266	12%	18,600	91%	383,739	31%	41,313
2010	100%	340,111	100%	840,000	100%	447,402	100%	335,338	45%	878,044	75%	121,268	100%	800,000	15%	210,221	100%	155,000	90%	378,086	54%	71,743
2011	100%	340,111	100%	840,000	100%	447,402	100%	335,338	80%	1,560,966	100%	161,690	100%	800,000	20%	280,295	100%	155,000	90%	380,688	75%	99,612
2012	100%	340,111	100%	840,000	100%	447,402	100%	335,338	40%	780,483	75%	121,268	50%	400,000	0%	0	100%	155,000	91%	382,628	38%	50,782
2013	100%	340,111	100%	840,000	75%	335,552	100%	202,645	20%	390,242	70%	113,183	62%	496,000	0%	0	100%	155,000	90%	381,645	31%	41,288
2014	40%	136,044	40%	336,000	0%	0	50%	101,323	0%	0	50%	80,845	0%	0	0%	0	55%	85,250				

1) Volumes shown are based on historic allocation percentages multiplied by current (2014) contract quantities for senior water rights and agricultural water service contractors. For M&I; volumes shown are based on historic allocation percentages multiplied by current (2014) historic use calculations (3-year average). This calculation method does not reflect the actual historic "allocated" volumes, but provides an illustration of what those volumes would be based on current contracts and M&I use rates.
2) The figures given for SRSC (Sacramento River Settlement Contractors) water is for Project Water only. In addition, Reclamation delivers approximately 1,775,508 acre-feet of "base supply" water representing the contractors' water rights water.
3) In 2009 only: the M&I Contractors on the American River received 100% of Contract Total, and those on the Sacramentio River received 75% of historic use.
4) Based on Historical Use of 202,645 acre-feet when Ag allocatin is less than 75%. If Ag is greater than 75%, allocation is based on contract entitlement and historical M&I use for contracts that do not specify Ag and M&I amounts.
5) Based on Historical Use of 161,690 acre-feet.
6) Friant Class 2 supplies represent a "supplemental" supply when water is available; Class 1 is a "firm" supply.
7) Uncontrolled Season
8) These figures are for water delivered to Project Contractors only and do not include an additonal 600,000 acre-feet delivered to satisfy water right entitlements of the Tri-Dams Project.
9) All data provided for refuge deliveries in Water Years 1993-1998 is based upon the best historical data available. Data for WY1999 forward is based upon refuge-wide uniform reporting system.
10) Water Year 2013 refuge delivery data is preliminary data.
11) Refuge figures shown are actual deliveries to refuges (both North and South of Delta), as opposed to allocations. Infrastructure and other limitations have historically limited ability to serve all refuge supply allocations.

Table 2 Central Valley Project South of Delta Agricultural Water Service Allocations and Hydrologic/Carryover Factors Affecting the Allocations

Year	Final Allocation[2 5]	October 1 Carryover Storage (TAF)[3]							Precipitation (in)[1]		Year Type[4]
		Shasta	Folsom	Trinity	Federal San Luis	New Melones	Millerton	Total CVP	Northern Sierra 8-station	San Joaquin 5-station	Sac Basin
1990	50%	2096.0	570.5	76.2	149.0	671.9	139.9	5003.5	36.0	27.7	C
1991	25%	1637.4	178.2	1162.4	387.8	377.7	183.0	3926.5	32.2	30.5	C
1992	25%	1339.9	506.1	670.2	268.6	296.3	174.7	3255.8	36.0	29.6	C

Table 2 Central Valley Project South of Delta Agricultural Water Service Allocations and Hydrologic/Carryover Factors Affecting the Allocations—Continued

Year	Final Allocation [2,5]	October 1 Carryover Storage (TAF) [3]							Precipitation (in) [1]		Year Type [4]
		Shasta	Folsom	Trinity	Federal San Luis	New Melones	Millerton	Total CVP	Northern Sierra 8-station	San Joaquin 5-station	Sac Basin
1993	50%	1683.2	171.6	838.3	95.9	83.8	164.6	3037.4	65.3	53.0	AN
1994	35%	3101.8	562.9	1947.7	347.2	670.7	179.4	6809.7	31.8	24.0	C
1995	100%	2101.6	216.9	1214.9	91.6	379.2	183.7	4187.9	85.4	69.6	W
1996	95%	3136.4	466.1	1872.6	442.4	1763.1	319.2	7999.8	61.3	43.5	W
1997	90%	3088.8	726.3	1712.7	177.4	1988.2	237.2	7930.6	68.8	54.4	W
1998	100%	2308.3	556.1	1493.9	131.8	1819.3	224.1	6533.5	82.4	65.3	W
1999	70%	3441.1	712.9	2077.3	713.0	2098.6	438.0	9480.9	54.8	37.0	W
2000	65%	3327.5	721.6	1961.7	110.6	1828.7	234.9	8185.0	56.7	42.0	AN
2001	49%	2985.1	660.7	1791.0	464.0	1803.6	211.9	7916.3	33.0	29.3	D
2002	70%	2199.6	367.6	1428.2	313.6	1481.0	187.9	5977.9	46.3	33.2	D
2003	75%	2558.2	509.8	1500.1	294.4	1278.2	221.1	6361.8	59.8	39.0	AN
2004	70%	3159.4	658.1	1881.2	286.5	1280.3	215.0	7480.5	47.3	28.8	BN
2005	85%	2182.9	376.4	1591.0	157.0	1110.1	181.4	5598.8	57.5	54.4	AN
2006	100%	3034.8	652.3	1991.3	402.4	1933.2	235.7	8249.7	80.2	55.1	W
2007	50%	3205.1	638.8	1973.9	402.1	2056.3	240.2	8516.4	37.2	24.9	D
2008	40%	1879.1	323.0	1550.3	194.3	1437.0	200.2	5583.9	35.0	27.9	C
2009	10%	1384.5	269.8	1461.1	36.9	1099.3	198.7	4450.3	46.9	38.9	D
2010	45%	1773.9	411.6	919.0	197.7	1108.4	350.4	4761.0	53.6	44.7	BN
2011	80%	3318.8	624.2	1557.7	374.3	1276.0	247.5	7398.5	72.7	65.4	W
2012	40%	3341.0	740.4	2166.8	642.1	2052.2	356.1	9298.6	41.6	25.0	BN
2013	20%	2591.6	451.6	1799.6	250.9	1510.7	318.5	6922.9	46.3	26.5	D
2014	0%	1906.0	361.1	1303.2	223.8	1047.1	317.1	5158.3			
Average	58%	2511.3	497.4	1569.7	286.2	1298.0	238.4	6401.0	52.8	40.4	

[1] Precipitation and year type columns are based on the water year (example 1990 represents 10/1/1989–9/30/1990).
[2] Allocation for CVP South of Delta Ag Water Service based on the contract year (example 1990 represents 3/1/1990–2/28/1991).
[3] Water in storage carried over from the previous water year (WY)—Example 1990 represents water carried over from WY 1989 that ended on 9/30/1989 into WY 1990 which starts 10/1/1989.
[4] Year types: C = Critical; D= Dry; BN = Below Normal; AN = Above Normal; W= Wet.
[5] 2014 allocation data is preliminary, based on initial allocations as of April 1, 2014.

Figure 4-1. Trends in Estimated Average Annual Delta Exports and SWP Table A Water Deliveries (Existing Conditions)

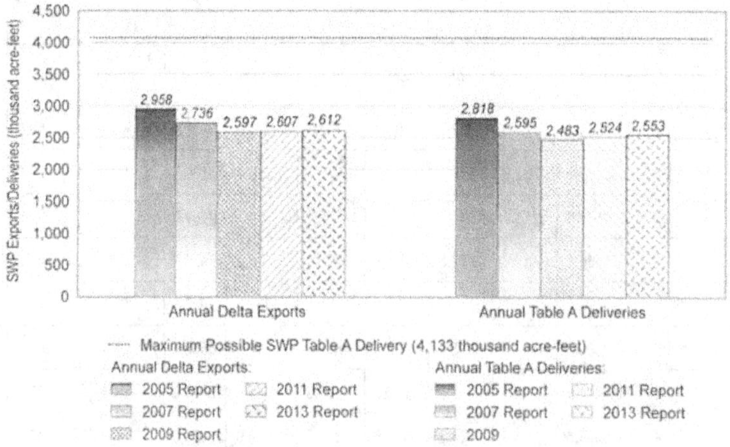

Figure SJR-20 **San Joaquin River Hydrologic Region Water Balance by Water Year, 2001-2010**

California's water resources vary significantly from year to year. Ten recent years show this variability for water use and water supply. Applied Water Use shows how water is applied to urban and agricultural sectors and dedicated to the environment and the Dedicated and Developed Water Supply shows where the water came from each year to meet those uses. Dedicated and Developed Water Supply does not include the approximately 125 million acre-feet (MAF) of statewide precipitation and inflow in an average year that either evaporates, are used by native vegetation, provides rainfall for agriculture and managed wetlands, or flow out of the state or to salt sinks like saline aquifers. Groundwater extraction includes annually about 2 MAF more groundwater used statewide than what naturally recharges – called groundwater overdraft. Overdraft is characterized by groundwater levels that decline over a period of years and never fully recover, even in wet years.

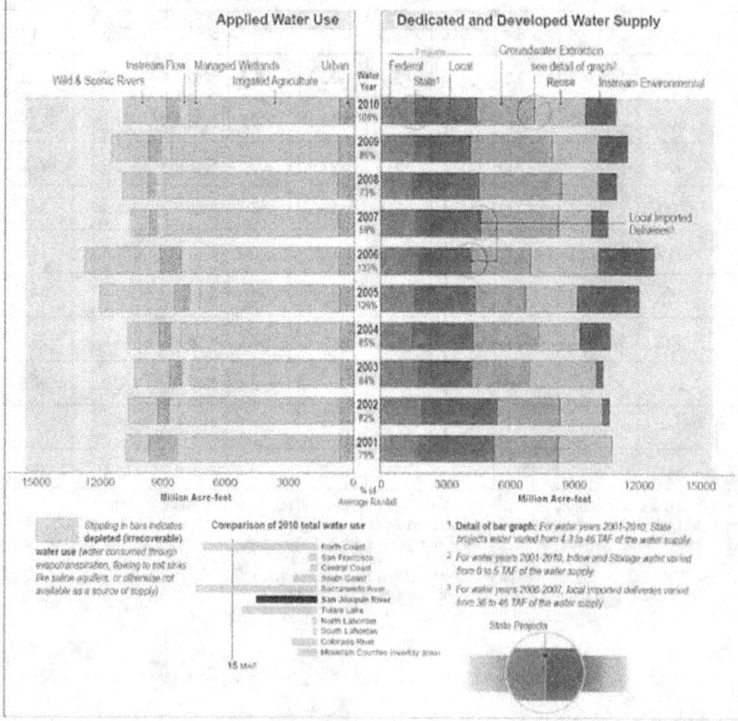

Key Water Supply and Water Use Definitions

Applied Water. The total amount of water that is diverted from any source to meet the demands of water users without adjusting for water that is depleted, returned to the developed supply or considered irrecoverable (see water balance figure).

Consumptive use is the amount of applied water used and no longer available as a source of supply. Applied water is greater than consumptive use because it includes consumptive use, reuse, and outflows.

Instream Environmental. Instream flows used only for environmental purposes.

Instream Flow. The use of water within its natural watercourse as specified in an agreement, water rights permit, court order, FERC license, etc.

Groundwater Extraction. An annual estimate of water withdrawn from banked, adjudicated, and unadjudicated groundwater basins.

Recycled Water. Municipal water which, as a result of treatment of waste, is suitable for a direct beneficial use or a controlled use that would not otherwise occur and is therefore considered a valuable resource.

Reused Water. The application of previously used water to meet a beneficial use, whether treated or not prior to the subsequent use.

Urban Water Use. The use of water for urban purposes, including residential, commercial, industrial, recreation, energy production, military, and institutional classes. The term is applied in the sense that it is a kind of use rather than a place of use.

Water Balance. An analysis of the total developed/dedicated supplies, uses, and operational characteristics for a region. It shows what water was applied to actual uses so that use equals supply.

San Joaquin River Water Balance by Water Year Data Table (MAF)

	2001 (79%)	2002 (82%)	2003 (84%)	2004 (85%)	2005 (120%)	2006 (133%)	2007 (59%)	2008 (73%)	2009 (86%)	2010 (106%)
Applied Water Use										
Urban	629	595	618	640	665	680	714	756	733	700
Irrigated Agriculture	7,243	7,612	6,998	7,505	6,559	6,982	8,124	8,177	7,809	7,045
Managed Wetlands	415	477	473	492	458	484	516	503	516	497
Req Delta Outflow	0	0	0	0	0	0	0	0	0	0
Instream Flow	1,424	583	600	582	772	1,046	361	345	614	644
Wild & Scenic R	1,091	1,420	1,714	1,504	3,611	3,557	883	1,232	1,755	2,090
Total Uses	**10,802**	**10,687**	**10,403**	**10,723**	**12,065**	**12,750**	**10,598**	**11,013**	**11,517**	**10,977**
Depleted Water Use [illegible]										
Urban	400	294	311	337	347	331	346	375	393	376
Irrigated Agriculture	4,938	5,605	5,279	5,687	4,922	5,485	6,304	6,515	6,221	5,421
Managed Wetlands	138	190	186	207	155	206	242	472	241	474
Req Delta Outflow	0	0	0	0	0	0	0	0	0	0
Instream Flow	0	323	318	304	335	553	96	0	0	0
Wild & Scenic R	0	797	0	1,123	2,555	2,077	532	708	1,044	1,184
Total Uses	**5,476**	**7,208**	**6,085**	**7,657**	**8,313**	**8,652**	**7,522**	**8,070**	**7,899**	**7,454**
Dedicated and Developed Water Supply										
Instream	0	323	318	1427	2890	2571	771	855	1358	1444
Local Projects	3,549	3,511	2,439	2,800	2,823	2,371	2,945	3,093	2,605	2,841
Local Imported Deliveries	0	0	0	0	0	36	46	0	0	0
Colorado Project	0	0	0	0	0	0	0	0	0	0
Federal Projects	1,764	1,904	1,765	1,461	1,542	1,738	1,640	1,453	1,472	1,552
State Project	4	9	17	14	5	7	24	10	46	30
Groundwater Extraction	2,969	2,930	2,688	3,073	2,351	2,814	3,604	3,853	3,846	2,709
Inflow & Storage	0	0	0	0	0	3	3	3	4	5
Reuse & Seepage	2,516	2,011	3,176	1,949	2,454	3,210	1,564	1,744	2,184	2,394
Recycled Water	2	0	0	0	0	2	2	2	2	1
Total Supplies	**10.802**	**10.687**	**10.403**	**10.723**	**12.065**	**12.750**	**10.598**	**11.013**	**11.517**	**10.976**

FIGURE TL—16 TULARE LAKE HYDROLOGIC REGION WATER BALANCE BY WATER YEAR, 2001—2010

California's water resources vary significantly from year to year. Ten recent years show this variability for water use and water supply. Applied Water Use shows how water is applied to urban and agricultural sectors and dedicated to the environment and the Dedicated and Developed Water Supply shows where the water came from each year to meet those uses. Dedicated and Developed Water Supply does not include the approximately 125 million acre-feet [MAF] of statewide precipitation and inflow in an average year that either evaporates, are used by native vegetation, provides rainfall for agriculture and managed wetlands, or flow out of the State or to salt sinks like saline aquifers. Groundwater extraction includes annually about 2 MAF more groundwater used statewide than what naturally recharges—called groundwater overdraft. Overdraft is characterized by groundwater levels that decline over a period of years and never fully recover, even in wet years.

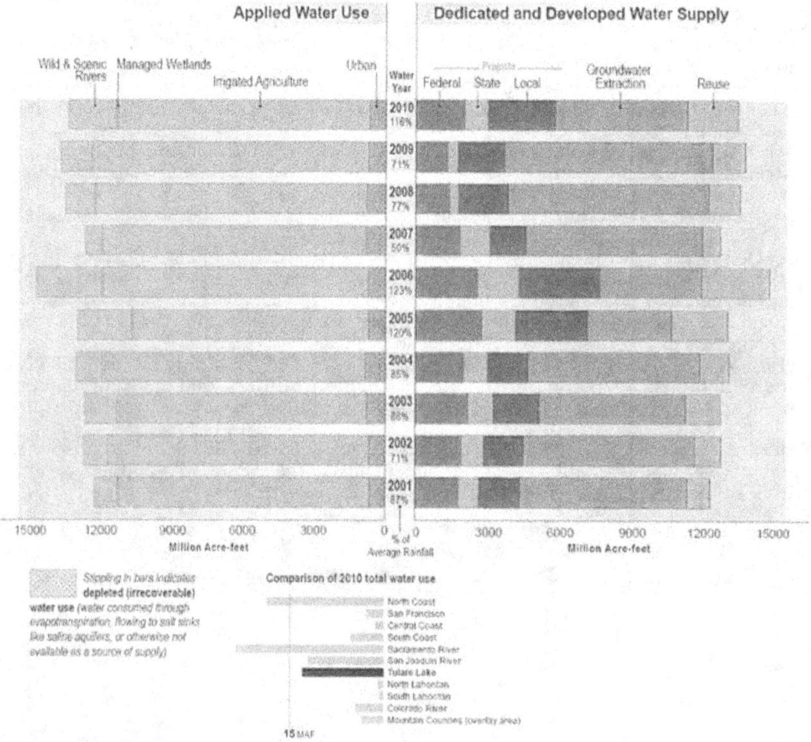

Key Water Supply and Water Use Definitions

Applied Water. The total amount of water that is diverted from any source to meet the demands of water users without adjusting for water that is depleted, returned to the developed supply or considered irrecoverable (see water balance figure).

Consumptive use is the amount of applied water used and no longer available as a source of supply. Applied water is greater than consumptive use because it includes consumptive use, reuse, and outflows.

Instream Environmental. Instream flows used only for environmental purposes.

Instream Flow. The use of water within its natural watercourse as specified in an agreement, water rights permit, court order, FERC license, etc.

Groundwater Extraction. An annual estimate of water withdrawn from banked, adjudicated, and unadjudicated groundwater basins.

Recycled Water. Municipal water which, as a result of treatment of waste, is suitable for a direct beneficial use or a controlled use that would not otherwise occur and is therefore considered a valuable resource.

Reused Water. The application of previously used water to meet a beneficial use, whether treated or not prior to the subsequent use.

Urban Water Use. The use of water for urban purposes, including residential, commercial, industrial, recreation, energy production, military, and institutional classes. The term is applied in the sense that it is a kind of use rather than a place of use.

Water Balance. An analysis of the total developed/dedicated supplies, uses, and operational characteristics for a region. It shows what water was applied to actual uses so that use equals supply.

Tulare Lake Water Balance by Water Year Data Table (MAF)

	2001 (85%)	2002 (71%)	2003 (86%)	2004 (85%)	2005 (120%)	2006 (123%)	2007 (50%)	2008 (77%)	2009 (71%)	2010 (116%)
Applied Water Use										
Urban	677	684	787	847	706	740	808	793	725	668
Irrigated Agriculture	10,567	10,917	10,437	11,008	9,944	11,195	11,150	11,439	11,668	10,663
Managed Wetlands	76	121	119	124	80	78	79	78	78	78
Req Delta Outflow	0	0	0	0	0	0	0	0	0	0
Instream Flow	0	0	0	0	0	0	0	0	0	0
Wild & Scenic R	964	1,019	1,387	1,099	2,285	2,757	668	1,228	1,264	2,017
Total Uses	**12,285**	**12,741**	**12,730**	**13,075**	**13,015**	**14,769**	**12,705**	**13,537**	**13,734**	**13,425**
Depleted Water Use (Mapping)										
Urban	246	240	286	303	246	253	271	277	245	228
Irrigated Agriculture	8,160	8,128	7,781	8,310	6,952	8,079	8,340	8,768	8,840	7,845
Managed Wetlands	39	62	58	60	55	49	59	60	56	51
Req Delta Outflow	0	0	0	0	0	0	0	0	0	0
Instream Flow	0	0	0	0	0	0	0	0	0	0
Wild & Scenic R	0	0	0	0	0	0	0	0	0	0
Total Uses	**8,444**	**8,429**	**8,124**	**8,678**	**7,253**	**8,381**	**8,670**	**9,105**	**9,140**	**8,124**
Dedicated and Developed Water Supply										
Instream	0	0	0	0	0	0	0	0	0	0
Local Projects	1,698	1,658	1,922	1,876	2,995	3,375	1,511	2,056	1,928	2,785
Local Imported Deliveries	0	0	0	0	0	0	0	0	0	0
Colorado Project	0	0	0	0	0	0	0	0	0	0
Federal Projects	1,788	1,896	2,175	1,977	2,749	2,575	1,849	1,403	1,314	2,021
State Project	849	948	1,046	1,021	1,404	1,727	1,223	377	434	979
Groundwater Extraction	6,985	7,144	6,120	7,187	3,504	4,253	7,371	8,397	8,711	5,537
Inflow & Storage	0	0	0	0	0	0	0	0	0	0
Reuse & Seepage	964	1,096	1,464	1,214	2,365	2,838	751	1,304	1,347	2,103
Recycled Water	0	0	0	0	0	0	0	0	0	0
Total Supplies	**12,285**	**12,741**	**12,730**	**13,075**	**13,016**	**14,769**	**12,705**	**13,537**	**13,734**	**13,425**

www.ingramcontent.com/pod-product-compliance
Lightning Source LLC
Chambersburg PA
CBHW080832180526
45168CB00006B/2652
* 9 7 8 1 5 1 2 3 4 0 5 9 4 *